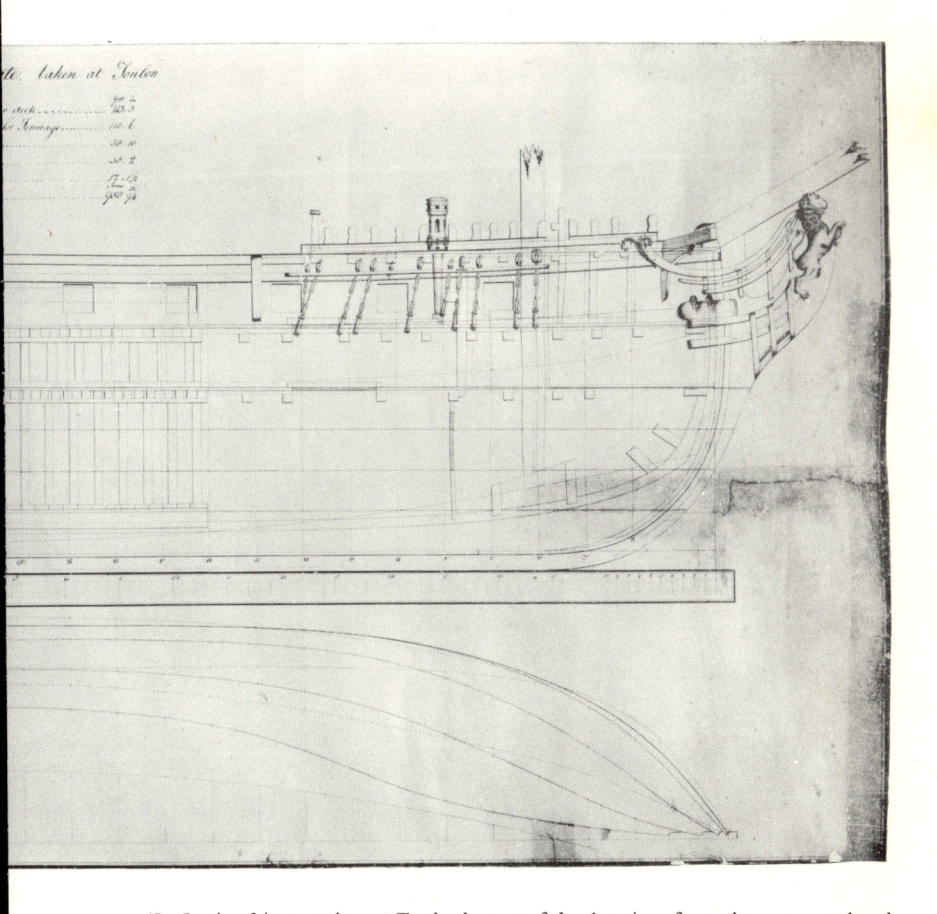

'La Lutine frigate, taken at Toulon'—one of the drawings from the survey ordered by the British Admiralty and now in the National Maritime Museum, Greenwich, London.

The Lutine Treasure

The Lutine Treasure

The 150-year search for gold in the wreck of the frigate *Lutine*

S J Van der Molen

English translation by James Brockway

Adlard Coles Limited   London

© Nijgh & Van Ditmar Rotterdam
This edition first published in 1970
and © Adlard Coles Limited
3 Upper James Street London W1
Printed in Great Britain by
Cox & Wyman Limited, London, Reading and Fakenham
ISBN 0 229 97482 1

Wedges of gold, great anchors, heaps of pearl,
Inestimable stones, unvalu'd jewels,
All scatter'd in the bottom of the sea

SHAKESPEARE: *King Richard III, Act 1, Scene 4*

The translator gratefully acknowledges
the cooperation of mevrouw J. Houwaard-Wood
on the translation of chapters
10 to 12 and 20 to 27

Contents

List of Illustrations	9
Foreword	11
1 Terschelling and the bullion ship	13
2 The frigate *Lutine*	16
3 How *La Lutine* became *The Lutine*	20
4 Some observations on the treasure	23
5 Aid for Hamburg	32
6 What the reports say	36
7 An eye-witness account	40
8 Some nautical observations	44
9 The tidal stream's timepiece	49
10 The first half-million guilders	55
11 Pierre Eschauzier fishes for a fortune	63
12 A royal decree and an association	64
13 Britons above the wreck	73
14 The men of Haarlem get a chance	77
15 The second half-million raised	81
16 W. H. ter Meulen in the Tease's thrall	93
17 The 'lamentable era' of the shell-fishers	102
18 Kinipple and Fletcher's British endeavour	108
19 England keeps the initiative	114
20 Van Hecking Colenbrander and Van der Wallen's coal-grab	119
21 Well-known salvors take the plunge	123
22 Beckers and his tower	127
23 The *Karimata* in action	139
24 A floating saucer	147
25 Lloyds and the *Lutine*	150
26 Relics of the *Lutine*	153
27 Numismatics and the treasure	157
Acknowledgements and bibliographical note	160
Notes	161
Index of Names	165

Illustrations

between pages 64 and 65

Frigate *La Dedaigneuse*, sister ship to the *Lutine*
Taurel's Dutch diver
Taurel's diving bell
Ter Meulen's sand-diving process
Lutine landmark on West Terschelling, 1863
Lutine landmark on Noordsvaarder sandbank, 1874
Lutine landmark on Vlieland, 1874

between pages 96 and 97

Ter Meulen's landmark slabs
Position of the wreck
The *Antagonist*
Model of Ter Meulen's sand-diver
The *Bill O' Malley*

between pages 128 and 129

The improved coal-grab
The *Texel* in action
The polyp-grab
Raising a cannon
Recovered relics
The first Beckers tower at Terschelling
Beckers' tower after being damaged
The second tower at Terschelling
The second tower at sea
The bucket-dredger *Karimata*
How *Karimata* operates
Van Capelle with 8lb gold bar
The *Texel* after recovery of gold bar
Coins recovered in 1930
The *Lutine* bell

Foreword

In 1799, on the dark and stormy night of 9 October, HMS *Lutine* was on her way to Hamburg from Yarmouth with a cargo of bullion. There seems to be some doubt as to its exact value, but most of the evidence points to its having exceeded £1,000,000. Caught on a lee shore in a NNW gale, the *Lutine* struggled to claw her way off the Friesian Islands, but the efforts of her master and crew were unavailing, and she struck a sandbank off the island of Terschelling, where she became a total loss. All but one of her 200 passengers and crew perished in the breaking seas on that wild night.

Needless to say, with a cargo of such importance to be salvaged, it was not long before attempts were made to reach the wreck. But the efforts were largely foiled by heavy silting as sand built up to great depths over the submerged hull. Although sunk in a mere three fathoms, the *Lutine* proved to be one of the most difficult of all treasure ships to reach. The efforts went on unremittingly until 1938 and different techniques were used, some of them most ingenious. Among these latter were ter Meulen's sand diver in the 1870s and Beckers' steel towers in the 1930s, neither of which enabled their inventors to partake of the treasure. In fact, in all this time, the only substantial amounts of gold and silver recovered were during 1800 and 1858 when a total of half a million guilders on each occasion were fished up. Nevertheless the constant lure of such a hoard waiting for a finder has been more than enough for enterprise after enterprise to enter the lists and join in the search for the 'gold in the waves'.

As an interested party, Lloyds of London have always been in the background. Their insurance covered the bullion cargo, but Lloyds never actually mounted a treasure hunt, leaving this to more speculative parties. However, as a reminder of the tragedy the *Lutine*'s bell which was recovered in 1858 now hangs in the Underwriters' room in Lloyds where it is rung on special occasions, once to tell of bad news, twice to tell of good. There is also a chair and table at Lloyds made from the rudder of the frigate which was brought to the surface in 1858.

This book tells the whole fascinating tale of the ship and her cargo. With great attention to accuracy the author assembles all the available information and then reconstructs the events as though they happened but yesterday. We are, indeed, taken again and again to the wreck to relive the experience of the salvors and those who risked their hard-earned savings in the enterprises. The best comparison is perhaps with a good detective story where the master detective unravels a complicated thread of evidence which leads to the final undoing of the criminal. The only difference is that here, sad to relate, there is no final arrest, no satisfactory conclusion in which the vast treasure is recovered, but instead there remains a question mark. Did the *Lutine* really carry that much bullion, and if she did, what became of it?

1 Terschelling and the bullion ship

A visitor to the island of Terschelling is bound to be confronted by reminders of an event in the annals of Western European shipping which still stirs the imagination, though many years have passed since the day the *Lutine* went to the bottom. If he lands from the ferryboat on the island's western shore, he will see two ancient cannon, one on either side of a massive anchor, pointing threateningly in his direction. Any inhabitant will be able to tell him that these cannon were among the many which the *Lutine* carried. But there is more to it than two bronze cannon. A stroll through the outskirts of West Terschelling, the island's capital, will take the tourist down Lutine Way (Lutineweg) and past a café which also takes its name from the celebrated bullion ship. The local museum, too, Het Behouden Huys, has preserved one or two items which were fished up from the wreck and which never fail to fascinate visitors.

If we are to believe one Gerben Colmjon, author of a book based in part of memories of his youth, the entire *Lutine* cult is only of recent origin. He writes rather scornfully that: 'It has become the fashion in recent years, when talking of wrecks, to start by mentioning that of the *Lutine*. In the good old days, no one on Terschelling ever gave it a thought. Even Meester Reedeker, who was always spinning yarns about the sea, never mentioned the name. It was just another ship that had been lost, although it was recorded in some book or other that hundreds of the victims had been buried near the lake in the dunes known as *Doodemanskisten* (dead men's coffins), where the coffins had been made for the corpses that were washed up. It was only in the autumn of 1910, when the very latest thing in modern British salvage vessels, with a large crew in impeccable uniforms and a half dozen officers who forthwith presented themselves to the British vice-consul, arrived off Terschelling to search for gold on the site of the wreck that one heard the name mentioned again, shrugged one's shoulders, and expected it all to come to nothing.'[1]

How different is the view taken by the late G. Knop, a philologist, historian, folklore expert and native of Terschelling, who recorded all his knowledge of the island and its history in a voluminous work entitled *Schylgeralân* – a rich source of authentic information. This author, born on Terschelling in 1873, had spent his youth there a quarter of a century before Colmjon's time, and writes that the *Lutine* 'would have been long forgotten with hundreds of other wrecks in the Schelling shallows if the lure of the glittering treasure which accompanied it to the bottom had not enticed generation after generation to attempt to lay hands on it.'[2]

There are other indications, too, that the *Lutine* occupied the minds of the island inhabitants more than Colmjon would have us believe. Knop prints a song by an anonymous author about the loss of the vessel, which had formed part of the repertoire of the pilots of former times. But more convincing still is a tale of the same period that Knop as a boy used to hear the pilots singing in their ships. J. A. Acket, in 1882, wrote, 'No wonder then that witticisms about the *Lutine* are always popular, while the sure knowledge that the treasure it contains lies there for the taking yet is, as it were, out of reach, lends it the appearance of a fairytale, which, in fact, it is not.'[3]

The author is right in remarking that the fact that attempts to recover the precious cargo of the *Lutine* attracted the greatest attention and helped to keep the memory of the bullion ship alive long after her grounding on 10 October 1799. For that matter, Acket knew a few islanders of great age whose recollections went as far back as the beginning of the nineteenth century, when gold and silver bars and coins to the value of over half a million guilders were recovered.

'The silver coins had turned black and it was thought that they would be of greater value if they were scoured clean, so that scouring went on for days on Terschelling.' The older inhabitants were still able to tell of 'this unusual work' and recounted 'all manner of anecdotes about the gold and silver fever which overheated the brains of the islanders so that at the time the island resembled a madhouse. The salvage of the gold and silver was sufficient, to cause the most sensible of men to lose their heads completely, especially if it is remembered that only the ordinary rate of salvage pay was paid for recovering objects of this value.'[4]

Shortly after 1882 a series of salvage attempts was begun in connection with the 'Hamburg vessel' – as the *Lutine* was sometime known – attempts which were to continue at varying intervals right up to the outbreak of war in 1939. Many inhabitants of the island were directly involved in these efforts, and if there was ever any risk of the beauteous ship with gold and chattels laden', as she was called in the old song, being forgotten, this salvage work, which often assumed quite spectacular proportions, was more than sufficient to keep her memory, and the hope of profit, alive.

2 The frigate *Lutine*

Little was known until recently about the sort of vessel the *Lutine* really was, other than that she was of the frigate type. However, the same name is not used everywhere to denote the same type of vessel, while in the course of centuries the type of vessel to which a name applies has changed too. In the nineteenth century the Dutch merchant navy understood by the term 'frigate' a sea-going vessel with three masts, all square-rigged. One spoke, accordingly, of ships which were 'frigate-rigged'. The term frigate did not have this meaning originally. In the beginning it indicated one of the smaller Arab vessels, scarcely bigger than a sloop. The type was not more than about fifty feet long and carried two sails and a number of oars. In the mid-seventeenth century the more elaborate version of this small Mediterranean galley disappeared, and later on the term was used for large ships-of-war which were finer of form and faster than the usual run of merchantmen.

During Holland's 'golden century', the frigate developed rapidly in size and armament. From a small vessel with six to twelve cannon, by about 1650 it had grown, in England, to be a warship with sixty-four cannon. During the eighteenth century frigates were increasingly important in the navies of the great European powers and also did service as privateers. Although the armament later increased still further, all the cannon were mounted on the same deck for some years.

The *Lutine*, which according to information from the Curator of the Musées de la Marine in Paris was launched at Toulon in 1779, originally had twenty-six guns, and it is not too difficult to form an impression of the vessel. In 1795 the *Lutine* had been rebuilt by the British Navy to convert her to a fifth rate frigate, 118 ft 6 in. between perpendiculars with 38 ft 10 in. beam and a depth of 12 ft 1 in. The length of the gun deck was 143 ft 3 in. and the burden 950 tons. The ratio of length to breadth was precisely 3:1.

There are no illustrations of the *Lutine* as she appeared at the time of the disaster, though drawings of the ship are to be found in the

National Maritime Museum at Greenwich. They were made after the vessel had fallen into British hands, to be measured and drawn without delay – an example of a rudimentary form of industrial espionage. By courtesy of the Curator they are reproduced here.

The *Lutine* was of round-bilge construction with three masts fitted with fighting-tops: the foremast and mainmast each in three sections and fitted with yards, while the mizzen mast, also in three sections, carried a spanker in addition to a square sail. Foresails were rigged on the bowsprit and jib-boom. Her sections were completely round, since she had a square, slightly projecting stern. Viewed broadside on, the vessel is seen to have two raised decks, a forecastle deck, or raised foredeck, extending aft of the foremast, and a quarterdeck. They were joined on the port and starboard sides by gangways, or catwalks. Companionways led down from these to the main deck. On the forecastle hung the ship's bell and between this bell and the foremast there was an anchor capstan extending through to the main deck, similar to the large capstan aft of the mainmast.

The *Lutine* was equipped with an eighteenth-century innovation: a steering-wheel instead of a tiller. A flight of steps led from the quarterdeck to the poop, the aft section of which consisted of the captain's cabin. This projected slightly, being supported by ornamented consoles, and had three windows port and starboard and nine (of nine panes each) at the stern. Carved columns were placed between the windows, but apart from these the *Lutine* was not rich in decorative carving, although the traditional lion rampant had been added at the bow.

As regards the vessel's armament, one always reads that the British increased the number of cannon from 26 to 32 and so far this has been accepted without question. But Mr J. D. A. Thompson, of Oxford, gave us a sight of his copy of the Navy List of 15 August 1799 (brought up to date to 1801), which had been drawn up by the Deputy Controller at the Navy Office. In this list the armament is recorded as follows: main deck, 26 twelve-pounders; quarterdeck, 4 twenty-four-pounder carronades and 4 six-pounders; foredeck, 2 twenty-four-pounder carronades and 2 six-pounders, making a total of 38 pieces, 6 more than the accepted number – 6 being exactly the number of carronades on board. In this list, in which the crew is shown as consisting 240 hands, the name 'The Lutine' has been struck through and the word 'lost' added – a feature which places the document's authenticity beyond doubt.

As far as the cannon are concerned, the muzzle-loading twelve-pounders raised from the wreck differ in length. The largest specimens are about 8 ft long, the smallest about 6 ft 6 in. The cannon on board the *Lutine* were mounted on wooden gun-carriages on wheels, which had come into use as early as the sixteenth century. One example of this type of gun has been recovered from the wreck. Each cannon was ranged with its barrel before a gun-port, which could be closed with a lid. These lids were raised when the vessel was made ready for combat.

Before the mainmast was the main hatch and forward of this lay the bomb bed, a second one lying immediately underneath it on the lower deck. This bomb bed was a stout timber frame, supporting the carronades.

Aft of the mainmast on the lower deck were the officers' cabins, five on the port and five on the starboard side, two on each side being forward of the partition which gave access to the messroom. From fore to aft on the starboard side ran the cabins of the ship's carpenter, whose job it was to maintain the hull, spars and ship's boats; the gunner, who was in charge of the guns and their use; the officer of Marines; the first lieutenant and the ship's surgeon. In the same order on the port side were the cabins of the boatswain, in charge of the rigging, gear, anchors, etc.; the captain's clerk, in charge of the ship's rolls and of keeping a check on the crew when in port; the sailing master, who, under the captain's supervision, was responsible for the navigation and the general supervision of navigational equipment, charts and instruments; the second lieutenant and the purser. The boatswain and the gunner each had a storeroom on either side of the foremast. The other main compartments which remain to be mentioned were a couple of 'bread bins' on either side of the stern, and a scuttle magazine (a scuttle is a small hatch in the deck) immediately aft of the mess room. There was also a 'lady's hole', of unknown significance, situated centrally between the two bread bins. It is a diminutive compartment, almost inaccessible and too small to be used as a store room or to enter, yet it strikes the eye clearly enough in the plans of old ships.

Descending via hatches one finally reached the 'after platform', that is to say a deck under the lower deck just mentioned, which extended forward from the stern over about a third of the ship's length. Here, right in the vessel's stern, were the two gunpowder stores, one either side of the light room, which was lit by oil lamps and separated from the gunpowder stores by a bulkhead with a round hole in it to let the light through on to the reflectors on the opposite walls – this with a view to the risk of fire – and, in addition, two bread bins (for bread or ship's biscuits and, properly, lined with metal against vermin) and two stores for spirituous liquors. Next to the hatch giving access to the hold below were the captain's pantry (starboard) and the first lieutenant's pantry (port). Finally, a well is shown on the drawing, forward of the after platform, against which the crate containing the shot was placed. If we also mention the hold, the ship's cellar, where still more store rooms lay and ballast was stowed, we shall have completed our survey. We have not seen the ordinary seamen's accommodation – they had their messroom in the forecastle and their quarters forward of the main deck. When the *Lutine* was on her way to Hamburg she had a few passengers on board. In vessels of her type it was the custom to accommodate such passengers on the quarterdeck.

The discovery of the drawings of the *Lutine* at Greenwich have made the ship seem to rise from under the waves again more than 160 years after she foundered. She is no longer a mystery ship, whose appearance we have to conjure up in the imagination, but now sails past us as she had sailed for years under the French flag before 1793, and as she sailed out of the Yarmouth Roads on that fateful morning of 9 October 1799.

3 How *La Lutine* became the *Lutine*

When the *Lutine* foundered in 1799 she was a British vessel, a King's vessel, in fact, which meant that she belonged to the British Navy. But the ship had not always sailed under the British flag. Launched in 1779, for fourteen years she had formed part of the French fleet, before falling into the hands of the British in 1793. It would be better, perhaps, to write that she was allowed to slip into the hands of the British, for she was captured without a fight.

To understand how the French should come to deliver up a French naval vessel to the British, it is necessary to know something about the political situation at the time. France had become a republic in 1792, but the political situation in that country was far from stable. All the same, after the Battle of Valmy, a great defensive victory over the Prussians, the French were able to undertake an ideological and military campaign against European despotism. A great part of the German Rhineland was captured, and at the same time Dumouriez drove the Austrians out of the South Netherlands. On 19 November 1792 a decree was promulgated offering fraternity and assistance to all peoples seeking to win freedom as the French people had done. On 31 January 1793, in his defence of the annexation of Belgium, Danton declared that France had a right to natural frontiers; in other words, to extend her territory to the Rhine, the Rhône and the Alps. On the same day Dumouriez was ordered to occupy Holland, while a day later, the Convention declared war on England, diplomatic relations with that country having been severed since the execution of Louis XVI. A few weeks later France was at war with the greater part of Europe; the First War of the Coalition, which was to go on until 1797, had begun.

The young Republic was now ringed by enemies – a fact which she was soon to notice. Dumouriez was obliged to withdraw before the Austrians and was defeated near Neerwinden in March 1793, going over to the enemy, who advanced into northern France as the Spaniards invaded the south. In the summer of that year the Federalist party (the Gironde) revolted and, although federalism was of little significance as a political idea, succeeded in stirring up such unrest in the French countryside that by mid-June about sixty departments were in more or less open revolt. In the south the main centres of this revolt were Toulouse, Bordeaux and Marseilles, followed by Nîmes and Toulon.

Although the movement came to nothing, the fact that the naval port of Toulon had sided with the rebels meant severe losses for France, viz, the destruction of the naval dockyards and of the squadron lying in the port. On 2 September 1793 the town had surrendered, or rather gone over to the English, after the rebels had entered into secret negotiations with them and the royalists in order, should the rebellion fail, to keep a back door open Royalist naval officers handed over sixteen French men-of-war to the British as part of the transaction.

In December, despite siege, shelling and bombardment by the French government troops, the British flag still fluttered above the town, but by the middle of the month the French forces, led by an unknown young man called Napoleon Bonaparte, had carried their penetration and infiltration so far that a considerable body of their men had occupied the mountain which dominates the town.

There was no other course left open to Admiral Hood than to relinquish Toulon, and he sailed away, taking ten or twelve thousand royalists in his ships. But before leaving, he had blown up thirty-four French ships of the line and taken twelve frigates with him to Gibraltar, all of which joined the British naval forces. Among them was *La Lutine* ('the tease' or 'tormentress') which, according to Steel's *Original and Correct List of the Navy* of July 1799, was taken to England in 1793 as a thirty-two gun frigate and became the *Lutine*. The British Admiralty did not omit to have the ship's measurements taken and drawings made – to which we owe the precise plans of the vessel – and possibly as early as 1794, and certainly not later than 1795, she was entirely resheathed in copper as far as the waterline, as was shown by the inscriptions hammered into these plates, found on remnants of the wreck.

```
MR
WARR: D
JAN 1795
28
```

```
MR
WARR: D
JAN 1795
32
```

The figures 28 and 32 appear to refer to the weight in ounces of each square foot of copper. These plates were secured to the ship's hull with bronze or copper nails, and on these nails one can see what is popularly, but incorrectly, referred to as the 'Lutine marks', a kind of bird's claw or chicken's foot, known in English as the 'broad arrow' and found on government property. This mark, already familiar in the seventeenth century, occurs on other pieces of the ship's equipment, which indicates that they were added or altered at a British naval dockyard, probably Woolwich. According to Steel's *Navy List* the *Lutine* was still employed there as a guard ship in 1796 and was therefore, very likely, unrigged.

In January 1797, again according to Steel, the *Lutine* sailed out to do service in the North Sea under the command of Captain J. Monckton. In July 1799 Steel's *List* shows that the *Lutine* was lying at Yarmouth under the command of Lawrence Skynner (Lawrence was a mistake — the captain's christian name was, in fact, Lancelott. Her first lieutenant was Charles Aufrere and her ship's clerk Mr John Strong. Nothing else is known of the vessel's history under Captain Skynner's command, other than a note to the effect that when the packet boat *Prince of Orange* left Yarmouth on 2 October 1799 for a voyage to Cuxhaven with mails, passengers and a King's messenger on board, she was accompanied by the *Lutine*, which must have returned to Yarmouth within a few days, since it was on the morning of 9 October that she was to leave the same port again, this time on her last voyage.

ently convinced that the *Lutine* was definitely carrying pay for the troops in North Holland, a matter on which, however, there is no agreement, even though the head agent could point to a number of reports at the time of the disaster.

There can be no doubt that money was to be sent to the Netherlands for soldiers' pay.[7] On 27 September 1799 the Secretary of the Treasury wrote to the Lords of the Admiralty informing them that 'a sum of money in silver will be ready for transmission to Texel, for the use of the Army in Holland', and also mentioning 'a quantity of Bullion for Hamburgh'. The letter enquires if 'there can be got ready by next week or earlier, one or two ships to carry the same'.

The Lords of the Admiralty replied briefly that the *Amethyst* had been 'ordered to keep in readiness for the service required'. A letter from the Secretary of the Treasury to the Lords of the Admiralty dated 2 October followed, announcing that the 'silver coin for the army in Holland' had been delivered that morning on board the *Amethyst* at Gravesend but that 'the Bar silver for Hamburgh' would not be ready 'till next week'. The Secretary added that, 'it being important that the money for the army should be sent at once there was a necessity of getting a second vessel ready for the conveyance of the Bar Silver for Hamburgh'. Captain Cook of the *Amethyst* confirmed by letter of 3 October to the Lords of the Admiralty that the money for the army in Holland had been received by him, and in a letter of 23 October that he had sailed from the Nore on 6 October and arrived at Texel on the 9th, where he 'delivered the money for the army'. He 'left Texel again on the 20th and returned this day [the 23rd] to the Nore'.

It does not appear that the *Lutine* had to convey money for the troops as well, nor that the army got into financial difficulties as a result of the loss of this frigate. True enough, when from his headquarters at Schagerbrug on the notorious 9 October 1799 Field Marshal Frederick, Duke of York, writing to Henry Duncan, Secretary of State, mentions, besides the weather, 'the complete lack of all necessities' as one of the difficulties the army had to contend with, this could include lack of funds. But on the 20th he gives as one of the reasons for ending the expedition 'the great difficulty in this season of the year of securing supplies'. He makes no mention at all of the soldiers' pay, and by then the *Amethyst* had delivered the money.

Or had this vessel brought only part of the pay, the rest being in the *Lutine*? It is, of course, a powerful argument for money for the army being on board this ship that according to Vice-Admiral Mitchell's letter to the British Admiralty the sole survivor related 'that the *Lutine* left Yarmouth Roads on the morning of the 9th inst., bound for Texel, and that she had on board a considerable quantity of money'.

What was the *Lutine* doing at or near that island? The place where the *Lutine* foundered lay far east of Texel and this might indicate that the ship had not, after all, received orders to proceed to the waters around this island. But it is possible that with bad weather threatening, Captain Skynner did not consider it advisable to sail into the Marsdiep and consequently continued on his way. There would have been no hope either of sailing into the Vlie Passage. No vessel would attempt this with a storm blowing up from the NNW, since she would without any doubt be stranded on one of the sandbanks at low water.

Thus the presence on board the ill-fated vessel of army pay to the sum of £140,000 has not been substantiated. That the *Lutine* had gold and silver on board for Hamburg need not, however, be doubted. There is one point on which there is no certainty whatever, and this concerns a remarkable announcement in one of the London papers in March 1869. The *Nieuwe Rotterdamsche Courant* printed this announcement, after which it appeared in the *Leeuwarder Courant* of 2 April 1869: 'Moreover, the ship had the Dutch Crown Jewels on board, which the Prince of Orange had sent to England to be reset and polished, which work was done, as it is today, by Messrs Rundell and Bridges of Ludgate Hill, famous at the time as jewellers by appointment to the Crown. These precious jewels, hermetically sealed in a strong iron chest, had been taken on board the *Lutine* a few days before she sailed.' It was further reported that 'the Dutch government' had promised a reward of £8,000 'to have the Dutch jewels recovered, they being Treasury property'.

All this is far from clear. The term 'Prince of Orange' cannot have referred to William V, for he was still in England in 1799 and did not settle on the Continent until 1801. This report probably refers to the hereditary prince, who, regarded in those days as the hope of the House of Orange, had gone to the Continent in 1796, settling, in view of the Russo-British enterprise, in Lingen. He could therefore have sent the Crown Jewels out of Germany to England via Hamburg, having them returned, after repairs, via Hamburg again. But what, then, was this Dutch government, which had made £8,000 sterling available to have the jewels recovered as being 'Treasury property'? If by 'recover', salvage from the wreck was meant, one might suppose that the Batavian Republic (the name of the Netherlands at the time) had made this sum available in the knowledge that the jewels were worth far more. But even then questions remain. How, for instance, did the Dutch government in The Hague know that the precious chest had been shipped on the *Lutine*?

This recondite detail is, for that matter, in keeping with the romantic story the anonymous author had to tell. He was also able to relate: 'As regards the ship's departure, it is also reported that the Captain was so impressed by the important commissions he had received from the bankers that on the eve of the ill-fated voyage he invited the entire elite of Lowestoft and Yarmouth to a ball on board ship.' This is the most appropriate place to point out that mention of this ball occurs again in a letter which E. B. Merriman sent to the Committee of Lloyds. This letter is reprinted in a memorandum about the *Lutine* which Sir H. Hozier, Secretary at Lloyds, wrote and which appeared in print towards the end of the last century.

But Merriman must have got his information from hearsay, which explains why he not only speaks of the ball, but also has the following hardly credible 'latest particulars' about the ship to report. 'The merchants interested in the fate of the crew and passengers learned, as a result of enquiry, that the last that was seen of the *Lutine* was about midnight, or later, of a clear night by the crew of a fishing boat that passed her quite close, and were attracted by the fact that on board the King's ship, instead of silence and all lights out, there was brilliant light, and evidently much joviality in the State cabin' (Hozier).

Even had a festive spirit reigned on board the *Lutine* – and the presence on board of young and rich passengers makes this feasible – it would still be difficult to believe that the weather changed so suddenly that a few hours later the ship was lost.

In the same London newspaper article of March 1869 it is also claimed that the *Lutine*, 'en route for a port in the Zuyder Zee', had money on board 'intended as pay for the English troops'. But as regards the amount of this pay the author has his own ideas, speaking of £150,000, in other words, £10,000 more than the usual figure quoted. This brings up the possibility that some gold and silver was recovered clandestinely.

It has been asked before, of course, whether any successful 'fishing' had been done, apart from the official salvage attempts. In 1861, Brand Eschauzier[8] makes some mention of this after having declared in a note that there were no grounds for assuming that 'more had been recovered at various times and on various occasions, officially or unofficially, than salvage to a value of two million guilders'. People had wondered now and then whether there was also 'any question of unofficial (that is, clandestine) salvaging', but the answer in the author's view was obvious.

'To anyone acquainted with the position of the *Lutine* in relation to the surrounding inlets and navigable channels, to anyone who knows the diligent and unceasing watch the sea-faring inhabitants of Terschelling keep on these sea-lanes and, finally, to anyone aware of the problems attached to fishing for *Lutine* treasure, any clandestine activity such as is supposed will be a complete impossibility.' In support of this view, Brand Eschauzier mentions that in 1858 four silver bars were recovered which completed the series of numbers on the bars salvaged in 1801, the same applying to the gold bars. The fear that the numbers missing in 1801 suggested that bars had been clandestinely removed is belied by these facts, as the writer justifiably remarked.

He also gives great weight to the fact that in 1860, when the helmeted divers from Egmond threw in their hand, Lloyds Committee went to the expense of sending 'an excellent diver's apparatus with an airpump and accessories' from London, also sending the Britisher, Heinecke, with two experienced English divers, to Terschelling to make a personal inspection of the wreck. If there were not 'still considerable sums of money' present in the wreck, Brand Eschauzier thought Lloyds would not have gone to such expense – to which argument one could reply that, as stated above, the Lloyds agent in Amsterdam had shortly before (1858) calculated the total value of the cargo to be £1,200,000, ignoring the possibility that some of this may have disappeared clandestinely or the question of whether this money was conveyed to Hamburg in one vessel only, the *Lutine*. Moreover, would they have been so well informed in London about what had been happening to and around the *Lutine* since 1799?

That there had indeed been poachers around the wreck before the official salvage operations began, we learn from a statement of 28 November 1799, by Robbé, Chief Receiver of Wrecks, that vessels were above the wreck almost daily 'to haul up anything they fancied'. However, this need not signify that it was gold and silver these uninvited guests managed to salvage. If this gold and silver had been loaded in the vessel's peak, it would have been no easy task getting it out, for much very heavy and cumbersome cargo would certainly have been loaded on top in this compartment and would have had to be removed before they could delve any deeper. This was still the position as late as August 1800 when, according to Robbé, a considerable quantity of heavy quality eighteen inch cut rope was raised, and there was still a lot of rope that would have first to be removed if they were to continue working. It is difficult to imagine that clandestine 'fishers for gold' did not encounter difficulty with obstacles of this kind. When the official salvage work was at last under way, adequate measures were taken to deal with unofficial competition of any real significance.

There is, finally, one point which should be gone into here. This concerns the information which can be deduced from the numbers on the gold and silver bars recovered. In the Netherlands it had been pointed out in the last century that there were five series of silver bars marked with varying letters and that the G series went up to the 100, AL went up to 69, SS to 63, the MS series bore the figures 1, 2 and 3, and the SSC series did not go beyond 96. It was deduced from this that each series went up to 100 and that the missing numbers must still be in the *Lutine*. The gold bars also bore letters and figures. In 17 cases the highest figures ran from 10 to 99, but in the M series the figure of 391 was found and in the SL series 447. Hozier, Secretary of Lloyds, had already pointed out in 1895 that if these figures are right, the idea of a system of series going up to 100 cannot apply. But in 1963 J. D. A. Thompson, a British numismatist, argued that at that time it was indeed the custom for gold bars to be arranged in sets numbered from 1 to 100. There is no information that silver bars were treated in the same way, but they do seem to have been treated in the same way in the *Lutine*. The presence of the marks M 391 and SL 447 among the gold bars does not, this expert thinks, rule out the possibility that 100 bars of each series were brought on board. (He was unaware, apparently, that in the W series, the numbers 128, 135, 136 and 137 were found.) The most striking example is provided by the series of silver bars marked with the letter G. All the bars from 71 to 100 are present in this series, with one exception, number 98, and a run numbered from 13 to 50 had also been recovered, again but for one bar, number 46. Below number 13 numbers 6 and 10 had been found. Taking the view that there were in the ship at least 18 series of gold bars, each of 100 numbers, and 5 of silver of the same number, this gives a total of 1,800 gold and 500 silver bars. In 1900 W. R. Kinipple made a similar calculation though he based it on 19 different series of gold and 5 of silver bars, making a total loaded on to the vessel of 1,900 and 500.

Their value at the time amounted to £1,320,000, in addition to which was the gold and silver coin.

Going over the reports of the number of bars salvaged, one cannot escape the impression that here, too, little attention has been paid to accuracy. This is particularly true of the results of the salvage attempts of 1857–59. There is agreement regarding the number of bars recovered in the 1800–01 period: all sources record 58 gold and 35 silver bars. In his *History of Lloyds* (1876), Martin quotes the same figure, but he records 86 gold and 97 silver recovered in all, which figures leave only 28 and 62 for the 1857–59 period. Hozier quoted almost the same figures in 1895; 85 gold and 97 silver in all. Kinipple (1900) arrived at the figures of 100 gold and 99 silver bars.

We do not know what these British experts' sources were. It certainly was not the printed report (1861) on the second successful phase of the salvage operations by Brand Eschauzier, *Nadere mededeling* (Further Information), which precisely recorded the outcome of each day's efforts. Martin and Hozier could have been acquainted with this publication; Kinipple, perhaps, used it. Be this as it may, according to Eschauzier's records 40 gold and 64 silver bars were raised at the time, which, added to those salvaged in 1800–01, totals 98 and 99.

The matter becomes more complicated still when we study the list Dros and Doeksen included in a pamphlet (1928) in support of their projects.[9] Here, too, one finds every bar mentioned, with mark, number and date, from each period of salvage, but totalling 76 gold and 64 silver bars. Since Brand Eschauzier mentions marks but no numbers, we have had to try to solve the riddle by comparing the day to day finds. The riddle, however, could not be solved: in six cases the figures of 1861 and 1928 did not tally.

Before we add this riddle to all the others that the loss of the *Lutine* left unsolved, we must add that J. D. A. Thompson is inclined to regard the letter marks on the bars as the initials of the financial organizations which dispatched them. Thus M would stand for the Royal Mint, BB for Barclay's Bank, HB for Hoare's Bank, and P & C for Parish and Co. To this one could add the G for Goldsmid, while further enquiry into the names of banks in London in 1789, in particular, Lombard Street, would no doubt yield further combinations of letters.

Finally, those concerned with this matter who feel that nothing is 'as safe as the bank', should know that according to Hozier, when asked for its opinion in 1894 in connection with the new salvage operations then being planned, the reliable Bank of England 'thought it reasonable to suppose that the intervening bars between the lowest and the highest of each of the lettered series that had been found, were still on board'.

5 Aid for Hamburg

When the Dutchman Vincent Nolte published his recollections as a former merchant, he did not omit to mention the calamitous voyage of the *Lutine* and the mission on which the vessel had set sail from England. He wrote: 'When the Hamburg Stock Exchange was in dire straits in 1799, the Stock Exchange in London had endeavoured to come to its aid by sending ready money. To this end they had acquired from the government the use of the frigate *Lutine* which took on board silver to the value of £1 million (more than 12 million guilders) and set sail for Texel. There is no need for me to describe the eager anticipation with which the ship's safe arrival was awaited. It is as easy to imagine as is the disappointment which followed when the grievous news was received that the vessel had gone aground off the Dutch coast near Texel and had been lost with all hands; if I am not mistaken, the second mate was the sole survivor and the sorry bearer of these tidings.'[10]

Despite certain inaccuracies – Nolte mentions only the silver on board and says the ship went aground off Texel – this recollection does at least make us aware of the immediate purpose of the voyage, which was to get trade moving again and to meet the financial crisis which was claiming so many victims in Hamburg and elsewhere.

Both German[11] and British sources tell us that the city on the Elbe had expanded rapidly in population and as a centre of trade. The *True Briton* of October 1799 put it briefly and concisely: Hamburg, lacking factories and production and known solely for being a large staple market (while its commercial activity was restricted to agencies, export and some foreign exchange transactions) was suddenly transformed by the French Revolution into a commercial centre of the first rank, a competitor of Antwerp, Rotterdam and Amsterdam. 'Hence Hamburg – became at once the representative of several other capital marts, without the proper means or the necessary circulating medium for such a wonderful increase of commerce. The trade of the city was no longer confined to European ports, nor to the produce of European manufacturers, but extended itself to America, the East and West Indies, etc.'

The biggest successes were, however, recorded on the money market. The last decade of the eighteenth century witnessed the city's astonishingly rapid evolution as the main banking and financial centre of Europe, a position that Amsterdam had been forced to yield. The volume of trade in currency, bills of exchange and money transfers rose by a tremendous amount, partly as a result of Hamburg dealers constantly increasing their personal contacts with dealers abroad. In these years Hamburg was also recognized all over Europe as the city doing the most business in insurance.

Dangers were lurking, however, even though the majority of the merchants were blind to them. They lurked not only in the uncertain political situation but also, and particularly, in the fact that things had developed so rapidly and threatened soon to get out of control. The period was characterized by almost unlimited demand for money, and since forward buying was still unknown on the Hamburg Stock Exchange, all goods bought on speculation, for instance, had to be kept and paid for in ready cash, should they not already have been sold for profit before the date of delivery. Endeavours were made to obtain the sums of money required by means of bills of exchange, which often involved what is known as 'kite-flying' on a large scale.

In 1799 stagnation set in. Consumption declined everywhere, goods piled up, prices fell sharply, money was short and reliable bills of exchange difficult to come by. Resort was now had to bad bills (from dubious debtors) and kite-flying increased still further. The result was a crisis with countless casualties.

During this period members of the English court threw English cotton textiles and English linen in enormous quantities on to the Hamburg market at bargain prices, capturing the market in north and central Germany from there.[12] This advance of English trade went hand in hand with a growing appreciation of the English element in commercial circles. In 1800 reporters told how there was a ludicrous preference in Hamburg for everything that was English.

In England itself, however, they were already having a lean time of it. The harvest of 1799 was a very bad one. The need to import grain called for exports to balance this. In 1800 the country was obliged to pay £23¼ million to the Continent, much of it to finance the increased purchase of grain. This payment could be made only by surrendering gold and the Bank of England's reserves accordingly shrank rapidly.

In this situation of crisis, attempts were naturally made to stem the avalanche. An item in the *St James Chronicle* of 3-5 October referred to this, reporting that 'the Bank of England on Thursday [3 October] came to the resolution of lending assistance to the merchants to the amount of one million and a half', which announcement stood in direct relation to the report dated Hamburg, 6 October, in various English newspapers[13] to the effect that:

> The following letter dated London, 3 October, from Goldsmid and d'Eliason, has been received by Parish & Co. It proves the efforts made by the merchants of London to support the credit of the foreign houses.
>
> "It gives us the greatest pleasure to inform you that the Bank of England has given us permission to export a great sum in silver and gold, which we shall send off by the packets of Sunday 6 October and Friday 11 October to Cuxhaven. We entertain no doubt that this seasonable relief [loans] will restore public credit. As soon as it was known here that our application was granted, several other houses made similar ones, which have had similar success. The good effects of these measures are obvious, and we hope soon to see the course of the Exchange again at 36." '

The gold on board the *Lutine* could not have contributed to that recovery. That gold and silver for Hamburg was on board is beyond dispute. There are letters and reports that clearly point to it. In the first place, there is an order, dated 9 October and addressed to Admiral Lord Duncan, who had the North Sea Fleet (for the time being at anchor in Yarmouth Roads) under his command, 'to send a Brig or Cutter to Gravesend for the service of receiving on board some bullion and conveying it to the Elbe'. On 11 October the Admiral replied from on board the *Kent*, stating he had ordered Lieutenant Wood of the *Nile*, an armed cutter, to proceed to Gravesend and place himself at the orders of the Treasury. In the same letter Admiral Lord Duncan informed the Admiralty that he would be sending a second warship to the Elbe: 'having received yesterday a pressing application from the merchants to convey a quantity of bullion lying here to Cuxhaven, I have ordered Lieutenant Terrel of the *Courier,* armed cutter, to proceed thither with it'. Lieutenant Wood, in command of the *Nile,* arrived at Gravesend on 12 October, informing the Admiralty of his arrival that same day. On the 14th, at the moment he hoisted sail to proceed to

Cuxhaven, he also sent a message saying he had 'received on board the bullion from the house of Messrs. Goldsmids and Co.' and, 'there being no post today from London, I have judged it good for His Majesty's service to proceed with the said bullion at once to the place of its consignment'. The *Nile* had a good run, although this did not prevent Wood from receiving a sharp reprimand for having left Gravesend without definite permission from the Admiralty.

As the ensuing correspondence shows, Admiral Lord Duncan did not carry out his first intention of meeting the wishes of the merchants by sending the *Courier* under Lieutenant Terrel to the Elbe with the gold bullion. Martin considers it at least 'highly probable' that the quantity of bar gold and silver mentioned in Lord Duncan's letter of 11 October was very much larger than had originally been expected. It is not improbable that additional amounts had been added once it was clear that the treasure 'would be in the safe custody of a man-of-war'. In any case, the Admiral, as Martin reports, discarded the cutter, choosing the frigate *Lutine* in her place, 'one of the swiftest and best-manned vessel in his fleet'. The letter which shows this has been preserved. It was dispatched on 9 October by the Admiral from the *Kent*, lying in Yarmouth Roads, and read as follows: 'The merchants interested in making remittances to the Continent for the support of their credit, having made an application to me for a King's ship to carry over a considerable sum of money on account of there being no Paquet for that purpose, I complied with their request, and ordered the *Lutine* to Cuxhaven with the same, together with the mails lying here for want of conveyance, directing Captain Skynner to proceed to Stromness immediately after doing so...'

On 22 October the 'Committee for managing the concerns of Lloyds', in a letter addressed to the Admiralty, requested 'the favour of Mr Nepean to lay before the Lords Commissioners of the Admiralty the information that a sum of money, equal to that unfortunately lost in the *Lutine*, is going off this night for Hambro, and they trust their Lordships will direct such steps as they may think expedient for its protection to be taken'. The request was granted but not very graciously, writes Martin. Admiral Duncan was ordered to form a convoy, but was instructed at the same time to inform the merchants that they could not expect such protection for the packets a second time. This is the last letter in the Admiralty archives that refers to the *Lutine*.

6 What the reports say

The *Lutine* foundered on the night of 9 October 1799 at about midnight. In our day a disaster of this kind would have been made known to the world the next morning, but in 1799 the news took longer to reach the public. Messages had first to be conveyed from the islands to the mainland by ship, after which it still took a long time for the postchaise or a man on horseback to get the information to the newspapermen. Speedy reporting of the news to the press in the case of the *Lutine* was further complicated by the fact that the Dutch coastline was kept under surveillance by the British Navy.

To obtain a proper picture of events we need to bring about some sort of order in the material available. We have no information about the actual time of departure. But if we are to believe a certain Warren R. Dawson, once honorary librarian of the Corporation of Lloyds, departure took place early in the morning. He writes that: 'The *Lutine* sailed from Yarmouth on her last voyage in the small hours of Wednesday 9 October 1799.'[14] A departure between midnight and five o'clock in the morning, that is to say, before light, would have been quite feasible; in the case of the *Lutine* the urgency of the mission may well have played a part. It may even be that the vessel left shortly after midnight. A report that 'the last that was seen of the *Lutine* was about midnight, or later, of a clear night by the crew of a fishing boat that passed her quite close, and were attracted by the fact that on board the King's ship, instead of silence with all lights out, there was brilliant light, and evidently much joviality in the State cabin',[15] implies that the vessels must have met in the early hours of the morning off the English coast. There is also a brief report of a sighting on 9 October after the crossing. Lieutenant James Anthony Gardner,[16] lying off Texel in H.M.S *Blonde*, reports that: 'A short time before we sailed we saw the *Lutine*, 36, Captain Launcelot Skynner, at the back of the Haaks, and if I am correct, the evening she was lost ...' As seamen will know, the Haaks sandbank lies west of Texel. Gardner does not mention the hour, although from his words it was evidently not yet evening.

On 19 October the British Admiralty received a letter of the 15th from Vice-Admiral Andrew Mitchell, commander of the actions against the Franco-Dutch (Batavian Republic) fleet, written on board *Isis*, 50 guns, lying in the Vlie Passage. The letter read as follows:

'It is with extreme concern that I enclose you herewith the copy of a letter I received the thirteenth inst. from Captain Portlock of H.M. Sloop *Arrow*, stating the total loss of H.M.S. *Lutine*, her officers and company, all excepting one man, on the outer bank of the Fly Island passage, on the night of the ninth inst.'

The letter enclosed, dated 10 October, was as follows:

'It is with extreme pain that I have to state to you the melancholy fate of H.M.S. *Lutine*, which ship ran on to the outer bank of the Fly Island passage on the night of the 9th inst. in a heavy gale of wind from the NNW, and I am much afraid that her crew with the exception of one man, who was saved on a part of the wreck, have perished. This man, when taken up, was almost exhausted. He is at present tolerably recovered, and relates that the *Lutine* left Yarmouth Roads on the morning of the 9th inst. bound for the Texel, and that she had on board a considerable quantity of money.

'The wind blowing strong from the NNW, and the lee tide coming on, rendered it impossible with Schowts [probably schuits, local fishing vessels] or other boats to go out to her aid until daylight in the morning, and at that time nothing was to be seen but parts of the wreck.

'I shall use every endeavour to save what I can from the wreck, but from the situation she is lying in, I am afraid little will be recovered.'

Various English papers printed similar accounts, in some cases with elaborations and some changes in fact from the official announcement. Nevertheless, it is strange that in both England and Holland the press should have published so little information. Even considering the fact that, owing to the British fleet's control of the seas, communication between the islands and the mainland was no easy matter, it is still surprising that a paper such as the *Leeuwarder Courant* should not have published the news until as late as 13 November, having received it from London via Paris (dated 25 October). After having mentioned the British seizure of the Spanish frigate *Thetis* with half a million piastres on board, the Dutch paper, quoting its British source, continues: 'On the other hand we have lost the frigate *Lutine* off the Fly Island, which besides a large number of passengers had a hundred and forty thousand pounds sterling on board for the Army at den Helder, after which it was to convey a considerable sum of money to Hamburg, to support the credit of the Trade in that town, so that the loss in money alone is put at five times a hundred thousand pounds; only one man of the Equipagie [sic] was saved.'

Neither in this issue nor in later issues is there a single line of news concerning the reactions to the disaster on Vlieland and Terschelling, where they would have had their hands full with the job of interring the washed-up corpses. Fletcher was able to collect a few more particulars about this sad occurrence from a statement made by a Captain Rotgans, J. T. Visser and a number of fishermen on 8 November 1857 that 87 bodies were found on the Noordsvaarder (then an exposed sand-bank out at sea, nowadays the most western point of Terschelling) and buried there. About 200 dead were laid to rest in a pit behind the Brandaris lighthouse, while three of the officers were buried in the churchyard by officers from British vessels lying at the time in the roads off Vlieland, and in a spot beneath three trees in the corner between the nave of the church and an outbuilding to the east. The island's *Dood Boek* (Register of Deaths) did not, however, record this interment, since no fee was paid. On a map Fletcher shows the place where the pit was dug near the Brandaris lighthouse as lying on the north side of the (former) graveyard of West Terschelling. It appears that rumour later connected the burial with a pool outside the village called d'Earmeskisten, which name was corrupted into the Dutch name 'Dodemanskisten'. Once 'Earmeskisten', meaning a place where the penniless or the corpses from shipwrecks were buried, was taken to be 'Doodemanskisten', or 'dead men's coffins', it naturally took very little persuasion to get people to associate this mysterious name with the dead of the equally mysterious *Lutine*.

But who were these dead? Only a few of their names are known to us: Captain Skynner, of a Navy family; First Lieutenant Charles Gastine Aufrere, only twenty-nine, yet already distinguished in his Navy career; John Strong, the steward; Walter Montgomery, ship's doctor.

The names of a few of the passengers are known, some from contemporary newspaper reports.[17] The Duc de Chatillon, son of the Duke of Luxembourg, was 'among the persons of note' on board. Another report says that the Duc de Montmorenci was on board also.

Merchants among the victims of the disaster included a nephew of the house of Goldsmid (apparently B. and N. Goldsmid of 5 Capel Court, London were referred to) 'who was going over with specie from the respectable house of his uncle's for the relief of the Hamburgh Merchants in consequence of the late failure'. From other sources[18] we know of the name of Daniel Weinholt of London who was taking a sum of £40,000 to the house of Parish and Co. at Hamburg. His body was washed up at Hornum and the Receiver of Wrecks had it sent to Rantum, placed in a coffin there and eventually interred in the churchyard at Westerland, on Sylt. A few years after this, Daniel's nephew, Arnold Weinholt, visited the island, had the church repaired – out of gratitude – and erected a marble memorial tablet to his uncle.

These, then, are the mere handful of names that history has handed down to us regarding the victims of the wreck. Their meagre number is entirely in keeping with the sparse information we have about the disaster.

The actual chain of events that led to the vessel foundering is a matter of conjecture. Brand Eschauzier, after pointing out that the fate of the *Lutine* was due to the treacherous Buitengronden sandbanks mentions the 'general feeling' that the vessel must have been thrown over one of these Buitengronden, thus sustaining a dangerous leak and as a result sinking on the Goudplaat. The total loss of all on board but for one survivor leads one, in any case, to suppose that the foundering was sudden and irremediable.

7 An eye-witness account

As we have already remarked, anyone investigating the *Lutine* disaster cannot but be surprised that so little information is to be found on the matter. Why did the handful of official papers that exist dispense with the drama in so few words? Why was no fuller report published in the papers? Did the disaster, the washing-up of hundreds of bodies and the burial of the victims fail to make any deep impression on the inhabitants of Vlieland and Terschelling, and was no report drawn up anywhere, no memoir written, no explanation offered by anyone at all? Fletcher went to some effort to assemble official data for his *Some Materials* but he did not find much. We had given up all hope of ever discovering any document containing details concerning the disaster from the mouths of contemporaries when sheer chance placed one in our hands, the text of which is given, word for word, for the first time in this present work.

Towards the end of July 1938, Mr. J. A. H. Reynders, the burgomaster on Terschelling, showed press reporters covering salvage operations by the *Karimata* 'a document yellowed with age, in which an eighty-year-old fisherman, one Folkert Visser, has recorded his experiences with regard to the *Lutine*'. When, from the particulars a newspaper made public about this document, we were able to conclude with certainty that it did indeed amount to an eye-witness account, we lost no time attempting to trace the document in the Terschelling municipal archives. It was here, too, that Mr. H. J. de Feyfer, Keeper of the Terschelling Musuem, Het Behouden Huis, finally located it, after it had lain for so many years in a box bearing the hardly promising label 'sundry documents received'. Although Fletcher had known of a statement by a Captain Rotgans, Visser and others, he had no knowledge of the existence of an actual document. The following English version is based on a Dutch text that has been slightly modernized for the sake of comprehensibility.

'In the year 1860, on the 14th March, there appeared before us, burgomaster of the island of Vlieland, one Jan Folkerts Visser, formerly a skipper by occupation, aged eighty years, resident on Vlieland, who stated that in the year 1799, on the 9th of October, at two o'clock in the morning, a sloop manned by twenty of the crew of the English corvette *Wolverine*, under the command of Captain Portlock,[19] arrived from the Vlieland waterway reporting that they had sighted rockets in the Vlieland Buitengronden and that a ship was in distress there. The skippers, pilots and seamen were ordered to make for the spot immediately in their boats, which order was obeyed with the utmost dispatch, nowithstanding the rough weather.

'That he, being present in person, the only surviving member (of those who were there at the time), with his father and six other persons went out in the pilot boat, that they put to sea at about five o'clock in the morning and there found many chattels and corpses, that they picked three corpses out of the water, one chest, a crate and two barrels containing flour, and afterwards came across a piece of the upper gun-deck of a ship, to which a man was clinging. That they thereupon immediately did all in their power to save him, that after much effort they succeeded in getting him aboard, after which they had returned to the waterway, since they saw nothing else other than all manner of goods and corpses. That they offered him some of the wine and bread (which they had taken from the sea), very little of which he used, and that he seemed to be out of his mind, seeing the commotion he made, none of which they could understand.

'That they arrived in the waterway at the British man-o'-war *Wolverine*, Captain Portlock, surrendering the three corpses, which were clad in uniform: blue coat, trousers and white waistcoat; one with two epaulettes, the others each with one. That after this the doctor of the said ship came on board their boat and gave the survivor medical attention, whereupon he swooned and was afterwards taken on board the warship, after which they returned to Vlieland and being on board the warship a day later, they then met the survivor, who was walking on a crutch since his leg was injured, and then learned that the survivor was the writer of the English frigate *Lutine* that had been in the command of Captain Lancelott Skynner; that they were then ordered to take the three corpses ashore on Vlieland, which corpses, so they understood, were those of the captain and two officers, which were buried in that place in the churchyard of the village of Vlieland.

'That thereafter Willem Blom, aged seventy-four years, appeared, who informed us among other things that he still remembered quite clearly how on the 11th of October 1799, the commander of the English frigate *Lutine* and two officers from the said ship were buried, that he had seen them in the coffin at the churchyard and had been present during the interment, while he has indicated on a drawing the place where the said captain and officers had been buried. That even some days after this many corpses from the said ship were washed up and buried.

'That they, appearing before us, have further jointly stated that the above is the truth, the genuine truth and nothing but the truth regarding all they had experienced in the year 1799 and which they still remembered clearly, after which they signed this document, together with us, burgomaster, on the day, month and year as stated heretofore.

Signed, Jan F. Visser, W. Blom, L. Zunderdorp, Burgomaster.'

We do not know why Burgomaster Zunderdorp had these elderly eyewitnesses make an official declaration. It may possibly be that this form of record was connected in some way with claims the burgomaster felt he could make to the *Lutine*'s cargo in his capacity as Vlieland's chief Receiver of Wrecks. As we shall see later on, in 1857 one of the members of the Eschauzier family (which possessed the exclusive right to organize salvage attempts) complained of the behaviour of the burgomaster of Vlieland, who first promised others 33 per cent of everything that should be brought to him from the *Lutine* and later joined with others in organizing a diving expedition. It is true that nothing came of this plan due to the intervention of a higher authority, but in 1860 Burgomaster Zunderdorp who, after all, was only a human being like everyone else, may have entertained hopes that things would change. This is no more than a supposition and the matter remains a mystery, likewise the fact that a copy of this document should have found its way into the municipal archives of Terschelling, unless the burgomaster of that island, in his capacity as Receiver of Wrecks there (and therefore being to some extent a rival of his colleague on Vlieland), had requested a copy or had had one sent to him.

The point is immaterial, anyway. Of far greater importance is what the old seaman, Jan Folkerts Visser, was able to conjure up from his remarkable memory there in the little town hall of East Vlieland (the building is still standing today, though it is used for other purposes) on 14 March 1860, supported later on, where some of the details were concerned, by the seventy-four-year-old Willem Blom. The story is related in sober terms and that is probably why it is so impressive and evocative. Who could fail to see in his imagination the actual décor against which the drama was enacted? A dark and stormy night, with the wind beating the sand of the dunes across the island and the thunder of the North Sea penetrating the little houses.

The manner in which Captain Portlock ordered the seamen on Vlieland to put to sea and go to the rescue is evidence enough that the British were masters of the seas in the region of the Wadden Islands where, only recently, on 30 August 1799, they had captured the fleet of the (Dutch) Batavian Republic, lying in the waters of the Vlieter, without having to fire a shot. What is more, the people living on Vlieland would not have been hostile towards them – all the island folk were suffering nothing but misery and loss as a result of a war into which they had been dragged by France; and although we have no information about Vlieland itself, it is not without significance that on the island of Texel, for instance, the British invasion had been greeted with joy.

The last sentence of the statement shows that numbers of the victims of the *Lutine* disaster found graves on Vlieland as well as Terschelling, even though the local register of burials, makes no mention of this. When one stands in the churchyard on Vlieland and reflects on the fact that in the shadow of a small group of trees at the NE corner of the small grey church (dating from the days of Holland's celebrated Admiral De Ruyter) some of the dead of the *Lutine* found a last resting-place, one is surprised to find no memorial at all to mark this mass grave. More surprising still is that inquiry should have shown that all memory of the events which Visser and Blom were still able to recall quite clearly a good century ago should have been extinguished among later generations. The little church of Vlieland, with its unique charm, in one corner of which the whalebones which once covered the graves of seamen who sailed for Greenland have been erected, is just the place for some reminder of the disaster of 1799, even if it were only a simple plaque. Yet far as we know, the only memorial tablet bearing a reminder of the unfortunate frigate is to be found in the village church at Easton-on-the Hill, near Stamford in Lincolnshire, where Captain Skynner's father was Rector.

8 Some nautical observations

In Holland, the North Sea coast is notorious for storms in the autumn. Unfortunately, there are no official weather records for the autumn of 1799, but known information seems to indicate that there was an early beginning to the usual period of gale force winds, reports from both British and Dutch sources making mention of 'angry' weather in October and November. Frederick, Duke of York, in a letter addressed from his headquarters at Schagerbrug to Secretary of State Henry Duncan, gave as one of the main difficulties hindering the invasion force 'the state of the weather'. This was on the same 9th of October that the *Lutine* had set sail. On 15 October he wrote to (French) General Brune of the possibility 'of waiting for the time when a change for the better in the weather' might lead to a resumption of the offensive, while on 20 October he reported that in North Holland the season 'had acquired the aspect of winter'. According to the Royal Netherlands Meteorological Institute in de Bilt, particulars from a source in Utrecht show that there was, in fact, a good deal of wind on that particular 9th of October. During November, too, the weather along the Dutch coast was generally bad. In a letter of the 7th addressed to the Commander of the Batavian Engineers, J. Goldberg, wrote that he expected that 'in this severe gale there must have been a large number of accidents along our coasts and in the Texel Roads and it must have been very unpleasant there'. Apparently there had been a continuous period of stormy weather, for Baron C. R. T. Krayenhoff reported to Goldberg from Den Helder less than a week later (on 13 November) that 'the daily storms' had resulted in a number of mishaps in the roads.

It is against this background that we should see the storm that to the *Lutine* signified her doom, a storm which apparently did not blow up until the evening. Neither at the moment of departure nor from the account of the sighting by Lieutenant Gardner in H.M.S. *Blonde* do we hear any mention of bad weather, and we may therefore assume, though with some caution, that during the greater part of her voyage the *Lutine* had a westerly to north-westerly wind. In the course of the afternoon or evening the wind increased to gale force, veering to NNW.

The tidal streams also have to be taken into consideration. Every seamen's manual warned – and still warns – sailors of the strong currents off the Dutch coast, currents which are much stronger when the wind is from the NW. The German *North Sea Manual (Southern Section)* contains the following advice: 'One should always set course well clear of Texel, for on the long stretch [from the Channel to the German Bight] north of the North Hinder lightship, no landmarks or coastlines are sighted by day nor by night any lights, and there are strong landward currents, especially when strong north-westerly winds are blowing and both the wind and the sea continually force the ship leewards. In such circumstances this advice is especially relevant to steamers carrying ballast at low speeds or sailing ships, which have even more leeway. What is more, north-westerly and northerly gales can increase the currents off these coasts to such a degree that it is advisable that sailing ships with the Dutch coast to the lee should not heave to in such circumstances but should steer the most favourable course and carry as much sail as possible in order not to be forced upon the coast.'

This is from the copy of the 1936 edition on board the Swedish motorship *Ecuador*, also lost on the Westergronden not far from the spot where the *Lutine* foundered. How, in the light of this, are we to explain Lieutenant Gardner's observations? If the *Lutine* was indeed seen in the vicinity of the Haaksgronden this could mean that she intended putting in at Texel, but that due to either the weather conditions or the approach of night, she turned north to await better weather or daylight. If we reject the Texel argument, it is conceivable that Captain Skynner wanted to do some reconnoitring, since Vice-Admiral Mitchell's fleet offered a degree of cover against possible French marauders, in which case the ship must have been set farther south than her captain had expected: a supposition hardly consistent with good seamanship.

Some general mention has already been made of the tidal streams, but to obtain a better understanding of what happened in the IJzergat, we need more concrete information about the situation at the time. For October the *Comptoir Almanak* reads: 'First Quarter, 6 October, 12.34 a.m.; Full Moon, 14 October, 6.07 p.m.'; while on 10 October the sun rose at 6.36 a.m. We assume that the moon passed the meridian about midnight on the night of 14–15 October, and since the moon's passage is roughly 48 minutes later each day we are able to calculate the time it crossed the meridian on 8 October as 7.40 p.m. Between the hour of the moon's passage and the high tide immediately following it, there is a certain time lag known as the lunitidal interval. This varies, although an average time can be calculated for all important ports and channels. The average time for Dover is 11 a.m. and on this basis we arrive at the approximate times of 6.30 a.m. and 7 p.m. for the high tide on 9 October. For example, the information in *Tides and Tidal Streams of the British Isles* (1909), referring to the situation before the damming-off of the Zuyder Zee, tells us that the tidal stream begins to enter the inlet between Vlieland and Terschelling from four hours and a quarter after high tide at Dover until three quarters of an hour before the next high tide there. So, on the evening of 9 October 1799, the stream began to set into this tidal inlet between 11 p.m. and midnight, when the *Lutine* was already in the greatest difficulty. What is more, an examination of British and Dutch current charts shows that in the area between the two extreme courses the *Lutine* may have been on, the tidal current was not in the vessel's favour from 1 p.m. onwards. If we add to this the NW to NNW wind, she must have gradually come into shallow water. It is impossible to indicate her route with any accuracy on the basis of the meagre information we have. Should the vessel have been in the vicinity of the Haaksgronden, she must have had the tidal stream against her during her last hours.

Should the *Lutine* have been making for Hamburg – that is to say, ignoring the possibility of her intention to call at Texel – Captain Skynner would have set a course well north of the Islands. On this NE to NNE course (they may have been trying to make the twenty fathom line) the ship would have had a fair wind, with the tide running north. During the afternoon, however, the weather worsened, the wind blew up, the sea grew rougher and from 1 p.m. onwards the tide, strengthened by the NW wind, had turned southwards. It may have been impossible to determine the vessel's position accurately and perhaps for a time the sounding lead, which would have been regularly employed, produced no results. Knowledge of the tidal streams in the North Sea in those days was minimal, which would have affected the accuracy of dead reckoning. So the thing of which seamen's manuals even in those days warned happened: the *Lutine* was driven towards the shore and with the Dutch coast to leeward her position was one of mounting danger. In the course of the evening she was borne down upon the lee shore, possibly without realizing her position, as the fires which were the primitive forerunners of lighthouses, had been put out owing to war conditions and the night was said to be 'exceedingly dark'.[20] If the information about the stranding taking place 'under a press of sail' is to be relied on, we can assume that the *Lutine* was carrying as much sail as possible and by sailing as close-hauled as she could, was still endeavouring to get out of shallow water.

Though it is not likely that she intended putting in to Texel, this possibility should be kept in mind. The possibility of the vessel having gone aground as a result of an unsuccessful endeavour to sail up the Vlie Roads can be ignored. As soon as the wind direction is north of west the sea becomes very rough, and as entering the Stortemelk channel under such weather conditions can be a risky undertaking even for the powerful motor vessels of today, despite guiding lights and gas buoys, we may safely assume that Captain Skynner would not have ventured to do this with his bullion ship.

Finally, something must be said about the position on the bottom in which the vessel was found by divers in the middle of the nineteenth century. Was it due to the chance interplay of various factors that the *Lutine* ran aground with her bows into the wind and current and accordingly did not capsize and break up? Or is this to be seen as the result of Captain Skynner's good seamanship? Running aground in a cross sea was considered the most unfavourable position a ship could get into. Being borne down upon a lee shore and anchoring as a last resort is a position which is anything but reassuring. Should one have an anchorage and sufficient crew to anchor with two, three or four anchors and to perform each manoeuvre separately and successfully, there is some possibility of saving ship and crew in this manner. This prompts us to ask whether anchors were used. When the wreck was inspected in 1858, two anchors were found in the bow of the *Lutine*, but it is not certain whether they were bower anchors. We know that in the period 1830–50 five to seven anchors were commonly used in British ships; viz., two day anchors, one sheet anchor, one emergency anchor, one bow or bower anchor, one heavy and one light kedge. If, therefore, it is not certain beyond all dispute that the two anchors were bower anchors, it is possible that during the last stage, before the vessel ran aground, one or more anchors were dropped. Unfortunately this, too, must remain an open question. What information we have does not warrant the verdict that there was negligence or faulty judgment – in short, that the captain failed in his task. We may not, therefore, attribute the wreck to an error of navigation.

9 The tidal stream's timepiece

The *Lutine* was not the first ship, nor was she to be the last, to fall prey to the treacherous sandbanks and currents for which the Dutch coast has always been notorious. There are no figures for the number of ships which were lost in the past either outside or inside the tidal channel or the Vlie, referred to by the hydraulic engineer J. H. van der Burgt as 'the most powerful of the tidal channels which penetrate the Dutch coastline'. Yet when it is known that in spite of the attention that has been paid from early times to marking this passage with buoys and beacons, no less than 11 ships went aground between 1876 and 1916, more than half of them being total losses, it is clear that in the course of centuries many hundreds of vessels must have been lost. In the period quoted, the sailing ship was, it should be remembered, being replaced in ever-increasing numbers by steamships, which are less affected by weather conditions, so that we must assume that the figures for the days when only sailing ships were used must have presented a very different picture.

The danger to shipping here has always been the presence of shallows (sandbanks, shoals) next to deep or fairly deep channels which – and this makes the situation even more precarious – are continually changing their shape and position. A study of charts of the area dating from roughly four centuries ago enables us to see what changes have come about in the sea bed of the Vlie and the tendency they reveal. On Terschelling the shallows lying off the coast which claimed the *Lutine* are called the 'Grongen' (a corruption of the Dutch word for shallows, 'Gronden') and they are fairly simple to distinguish by their position.

On an imaginary trip through the waters off Vlieland, the first channel we pass is the South Stortemelk, which runs along the north coast of the eastern stretch of the island. Immediately north of this waterway lie the Stortemelk shallows, and then sandbanks known as the Westergronden (western shallows) and the Noordwestgronden (north-western shallows), a group liable to marked changes in shape and, as a rule, penetrated by one or more channels. Beyond these come the Noordergronden, which are more or less permanently divided by a channel into two groups of sandbanks and are also divided from the Noordwestgronden by a channel, the position of which is fairly stable. The *Lutine* went aground in the Westergronden at a point indicated by Vice-Admiral Mitchell as 'the outer bank of the Fly Island passage' (by Fly Island he meant Vlieland), and by Robbé, Receiver of Wrecks, as 'in the IJzergat'. One will look in vain for this passage on a modern chart, for it has long since disappeared, and with it its name of IJzergat; it is only to be found on old charts.

Who would not imagine he had some reminder of the lost bullion ship before him when noticing on an 1831 map the name Goudplaat (gold shoals) indicating a large sandbank on the north-western edge of which the wreck of the *Lutine* has been indicated? Yet this name has nothing whatever to do with the bullion ship of 1799. It appears in the form of 'Goltplaat' on a chart drawn by N. Witsen in 1712 and entitled *Texel en Flie Stroom* (Texel and Vlie (Fly) Stream). By the nineteenth century it had been forgotten that 'goltplaat' and 'Engelse Hoek' had no connection with the *Lutine*.[21] We might ask ourselves whether another bullion ship foundered in the region at some earlier date, or yet another, carrying a cargo of iron, for 'Yzer' or 'IJzer' in Dutch means iron. 'IJzer-gat' does, in fact, appear on a chart drawn by Lieutenant A. A. Buijskes a few years before the *Lutine* went aground. This chart is entitled *Plan for entering the Vlie* and was 'commissioned by order of the committee on naval affairs'.

On this chart are shown three channels marked by buoys along which one could sail between the islands of Vlieland and Terschelling to or from the North Sea. From east to west they were the Boomkens or Schellinger Gat, the Russische (Russian) Gat and the Stortemelk (off the north coast of Vlieland). The IJzergat lies roughly halfway between the Stortemelk and the Russische Gat and is not marked by buoys. It appears on a chart by J. P. Asmus, dated 1786 and of very unreliable draughtsmanship. On this chart, entitled *Chart of the Narrows in the Vlie Inlets* ... the IJzergat is marked 'geen passage' (not a passage). It stands to reason, therefore, that when the *Lutine* entered shallow water at this spot she was bound to founder. Yet even had it been a through channel and marked by buoys, the *Lutine* would still have been unable to reach safety, but would certainly have met with the same fate at some other point in these shallows. In weather as rough as it was, there was no chance at all of a sailing ship like the *Lutine* being able to enter such an inlet.

The IJzergat also appears in an illustration of the passage between Vlieland and Terschelling drawn by the engineer A. F. Goudriaan in 1802, which chart is said to be in the collection of the Royal Institute of Engineers at The Hague. On this chart, to which an engineer, Mr J. H. van der Burgt, kindly drew our attention, the IJzergat is shaped like a bow-net formed by the Westwal to the south and an adjoining half-moon shaped region of shallows to the north, separated from the Noords-Vaarder by the Russe Gat. In 1824 the configuration had changed to such an extent, according to a chart by P. J. Grinwis, that the IJzergat had to be omitted.

The continual shifts in the pattern of the sandbanks and channels in the submerged delta of the Vlie played an important role in the attempt to salvage the *Lutine* treasure and we must therefore say something on the subject.

The mechanism involved in the effect of the tide on the inlets along the Dutch coast was explained by the engineer Joh. van Veen in 1936,[22] while in that same year another engineer, J. H. van der Burgt, described the changes in the sea bed of the Vlie, and in 1946 Volkert de Vries was in the same area to make a study of the geographical history of the western tip of Vlieland. It is interesting to read that it is to the tidal currents and the surf that Van Veen attributes the formation of the system of sandbanks and channels in the submerged delta of a tidal inlet. Two important factors are the direction of the tide and the direction of the prevailing wind. In addition, Van Veen says that the ebb tide exerts a far more powerful effect than the flood tide. He goes on: 'The wash of the waves, coming mainly from the west, shifts the sandbanks between the channels in an easterly direction. As a result, the channels also often move eastwards. The weaker eastern channels are the most readily forced aside, not only because they offer less resistance, but also because they lie in areas where there is a great deal of sand, and sand at a high level. The shifting of the sands brought about by the surf in the outer delta makes itself manifest by a concomitant and spasmodic swinging to and fro of the channels. Sandbanks shift eastwards and join up now and again with the western side (the head) of the eastern island.' The joining up of the Noordsvaarder sandbanks with the island of Terschelling in 1860 is a telling example of this process.

De Vries, who studied the situation near Vlieland, established that the surplus sand transported by the sea in the tidal inlet passes the various sandbanks with the hands of the clock, as it were, while temporary silting-up in the south-western outer delta region is always followed by a south-westerly breach in the main channel. The investigator found handsome proof of this in the behaviour of the well-known Stortemelk channel north of Vlieland. This navigable channel has silted up repeatedly as a result of the movement of a large quantity of sand along the island's beautiful coast, yet it has always opened up again. In 1680 it had become so shallow that it was no longer marked with buoys, and in 1710 it was completely silted up. However, by 1736 it was once again twenty-one feet deep and pilots were referring to it as a 'good and respectable inlet'.

It is interesting to see what W. H. ter Meulen had learned, though brought up far from sea and shore, from his own observations about the Vlie's tendencies to alter shape. He speaks of this regularly in his writings, since he was convinced that his system of salvage could only work if there were sufficient water at the site of the wreck, in other words if the quantity of sand on top of the *Lutine* were at a required minimum. In his report for the year 1874, he said that success was wholly dependent on 'the formation of a passage above the *Lutine*'.

'A passage to the sea to the sea has often existed in this place and it was casually referred to as the Noordwestgat; the inlet that bears that name today has been moving north little by little and has also been filling up with sand; the flood and ebb tides no longer flow through it so strongly as before and very often the water is full of eddies which are both the cause and the result of banks of sand and hollows: all this leads one to expect the current to penetrate elsewhere. As a rule, this happens at a point further south.'

Ter Meulen may not have been able to offer a satisfactory explanation (although he attempted to in his report for 1878), as Van Veen was to endeavour to at a much later date, but the phenomenon we are concerned with was well known to him; he even tried to give Nature a helping hand by means of what was known as a 'channel chain', which he put on exhibition in the Palace of Industry in Amsterdam in 1874. The inventor claimed that, when laid on sand in flowing water, such a chain caused a channel to form, although further tests would have to be made. Prince Hendrik of the Netherlands then expressed the wish that this chain should be tested out and Ter Meulen received a chain of 525 metres from the Admiralty for the purpose.

This chain was then laid on the *Lutine* sandbank 'in a position which corresponded as far as possible with the direction of the existing flood and ebb streams, i.e. roughly east to west, and deliberately slightly south of the *Lutine*'. But although the chain was provided with ballast and anchors, it shifted away, while of the 1,640 wooden planks riveted to it a hundred or two broke loose and floated away. When, however, strong clamps were used, no more trouble was encountered and the chain remained in position from 20 September 1874. On 20 October the last soundings were made, after which the swell, even in calm weather, was such as to rule out further soundings.

Ter Meulen considered it a drawback that the chain was comparatively short, had a large distance to cover and did not extend from deep water to deep water. All the same, as he wrote after that month, 'it seems to be breaking up the ridge of sand it is stretched over a little, and if this is indeed the case the breakthrough, that is to say, the formation of a passage above the *Lutine*, may shortly be expected'.

It appears that no one before him was aware that the sandbank in which the *Lutine* had sunk would sooner or later give way to a channel. The salvage contractors do not seem to have drawn any conclusions from the fact that only a few years after 1800 the first attempts at salvage had had to be abandoned because too much sand had deposited itself on the wreck. It was accordingly by mere chance that in 1857 it was discovered that the wreck lay uncovered. Had the contractors kept a watch on the situation in the inlet, it would have been possible to calculate roughly when the sandbank would be replaced by a channel. In 1831 and 1853 precise and reliable charts of the inlet were made on the basis of soundings and comparing them, Van der Burgt already found signs on the 1831 map which indicated that the main current, the Vlie Stream itself, was then seeking a new outlet to the sea. In 1796 this current flowed seaward in a gentle arc through Russische Gat in a north-westerly direction, but with the complete shift eastwards this channel silted up and the Nieuwe Gat came into being, running NE. But at the same time, due to the eastward advance of the Goudplaat the channel became constricted along the western flank of the Noordsvaarder and the ebb tide in the Vlie Stream tried to force a new breach in the outer banks in a westerly direction south of the Goudplaat.

From the chart for that year, Van der Burgt noted: 'In 1853 this breach in a WSW direction comes into existence, causing the Westergronden and the Goudplaat to join up to form the Westerbuitengronden' (western outer shallows). It was discovered four years later that near this breach a deep channel had formed, which stretched as far as the bank on which the *Lutine* had sunk. But this channel moved northward fairly quickly, so that as early as 1859 the wreck was covered with sand again.

Van der Burgt, however, on studying the charts made between 1830 and 1934, discovered something else which tied up completely with what Ter Meulen had anticipated. 'According to the marine charts there must have been a fairly shallow channel in the place concerned from 1915 to 1916, yet in those days salvage attempts had been abandoned owing to the war.' Although as a result an undoubtedly favourable opportunity had been let pass, in 1936 this hydraulics expert had not given up hope of the process repeating itself. In his study he writes: 'Now that the Boomkensdiep, the main channel followed by the Vlie Stream flowing northwards, is silting up, it is by no means improbable that a new channel will break through the Westergronden in a south-westerly direction. This channel will swing north and after the elapse of some years find its way across the wreck of the *Lutine*. When that happens, it will almost certainly be possible to recover the highly coveted cargo, thus bringing to an end a sensational episode in the history of salvaging off the coast of Holland.'

We do not know when this channel relieved the wreck of its layer of sand. Perhaps it was during the Second World War, which would mean that Mars had thrown a spanner in the works yet again. Or had the results obtained by the *Karimata* in 1938, during which attempt the entire wreck was probably pulverized, caused all serious interest in further salvage attempts on the floor of the Vlie inlet to dissipate, even under the most favourable conditions? Whatever the answer, the hands of the tidal stream clock have indicated the right moment more than once – but those moments were allowed to pass unused.

10 The first-half million guilders

Although Captain Portlock had not been very hopeful about the possibility of salvaging anything of value from the wreck of the *Lutine*, the Lords of the Admiralty replied to Vice-Admiral Mitchell on 29 October, stating their concern at this very unfortunate accident, and directing the Admiral to instruct Captain Portlock to take such measures as may be practicable for recovering the stores of the *Lutine* as well as the property on board, '*for the benefit of the persons to whom it belongs*'. The captain of the *Arrow* was further to be instructed 'to send over the man who had been saved, on the first opportunity, that such information may be given to the persons concerned in the property as may be necessary for the benefit of the Insurers'.

In the meantime, according to Martin, the underwriters had taken matters into their own hands almost a week previously, various persons having been sent out by Lloyds to see for themselves whether it would be possible to recover anything from the wreck. At the same time the underwriters promptly paid out in full on the entire loss. But Lloyds' agents soon returned empty handed. Meanwhile, the Dutch had not let the grass grow under their feet. The Netherlands were at war with Great Britain and so the wreck was declared to be war booty. Salvage attempts were also made. We now know that the old inhabitants of the island of Terschelling had not made up the stories which they told Acket in 1882 about wagonloads of gold and silver and the polishing of blackened coins. We are indebted for this information to Dr L. C. Suttorp, who published the results of an investigation into the official archives in the journal *Historia* (1935, pp. 209 et seq.). This important article does not seem to have attracted much attention, which is all the more reason for making grateful use of it for the following.

The year 1798 saw the setting up of the one and indivisible Batavian Republic. This had (on the French model?) a Committee for the Public Properties of Holland which was accordingly concerned with state properties, and especially with those exploited by the state for the sake of revenue. So it was this Committee which, soon after the news of the disaster, quickly took the necessary steps to make the war booty in question remunerative. It directed the Receivers of Wrecks on Texel, Vlieland and Terschelling to furnish information and, if possible, to procure more particulars. Twelve days went by before F. P. Robbé,[23] who was both sheriff and Receiver of Wrecks of Terschelling, received the letter. On 26 November 1799 he replied to the Committee and pointed out how improbable it was that any part of the wreck might be salvaged in the given circumstances, seeing that it was lying in an unfavourable position and that the season was far advanced. He had therefore come to the same conclusion as Lloyds' agents. Nevertheless, he did request that the monopoly for salvage operations might be granted to him for the near future: 'Should you be pleased to employ me in such a capacity, so that I, and none other, may employ several trustworthy persons, then you may be assured that I will use my utmost means, yea, leave nothing untried in order to recover everything possible when opportunity offers.'

He urged the necessity for haste, arguing that vessels were over the wreck almost daily to haul up whatever they could lay hands on. But Robbé had to exercise patience. It was not until 28 December that he recived the letter of authorization, dated 5 December, which gave him the sole right to carry out salvage operations, albeit with the qualification 'that the costs of this salvage work and in general of everything that is used for this purpose will not amount to more than a third of whatever may be recovered, either of the cargo or of the wreck'. Robbé did not waste any time; as early as New Year's Day 1800 he commissioned three salvors to go fishing and salvaging with crews selected by them and their own boats and equipment. They were promised a third part of the value of everything recovered, 'but all on the explicit understanding that not the smallest piece of property, whatever its nature, may be held back, but that o nthe contrary everything must be given up promptly'. Whoever defied this rule would not only forfeit all right to his own share but could count on 'the severest correction' prescribed by the law.

However, the salvors did have something to encourage them. They had no competition to fear, as 'power and authority' was granted to them 'to turn away each and every pilot or fisherman (whatever his place or origin) who causes them the slightest hindrance in their above salvage work by words or deeds, or who also attempts to recover goods from the wreck, near it or in its vicinity'. If these rivals did not choose to beat a retreat, 'strong measures' would be taken against the recalcitrants 'to prevent similar disagreeable incidents in the future'. In addition, on 28 May 1800 the government proclaimed a ban prohibiting private individuals from recovering and salvaging commodities out of the ship.[24]

But although the salvage operations were well organized on paper, Robbé and his salvors could not coerce nature. January 1800 went by without result; ice-drift in the harbour of Terschelling and heavy surf outside it were serious hindrances. Neither was spring favourable to the undertaking. It was not until August that the Committee received enthusiastic reports from Terschelling: they had not only reached the wreck but had also wrested various objects from it, such as 'a considerable quantity of hacked heavy 18 in. cordage and inferior sorts which, although lying under water for a long time and being very tangled, appears to be in very good condition and can best be used at short notice; furthermore four iron pigs of ballast have come out, and a considerable number of cannon, cannonballs, etc., can be seen when the water is clear, but a great amount of cordage must be removed from it first before one can get at it'.[25] A week later the wreck also yeilded up the first treasures: an open cask with seven gold bars weighing 81 pounds in all and a small chest with 4,606 Spanish piastres.

How did the inhabitants of Terschelling react to this valuable find? Robbé gives a faithful account of it in his letter to the Committee: 'This has caused the animosity of a good many fishermen to increase by several degrees, and everyone tries to persuade me to engage him too, which cannot be done by any means.' Who, in reading these lines, does not see Robbé pacing along the shelving, unpaved, sandy little streets of the township of West Terschelling, being halted time and time again by envious fishermen who wanted to have a hand in the salvage work too?

But although the Receiver of Wrecks was capable of keeping his own people under his thumb, he very soon had to report trouble with some fishermen from Urk, who had taken some cordage by force, though they had had to give it back later. Apparently the inhabitants of this islet in the Zuyder Zee, who used to fish in the North Sea in those days, made such a nuisance of themselves that Robbé had to acknowledge that he would be better off with them as friends than as enemies. And so he came to an agreement with his opposite number on Urk whereby the latter would send 'twenty of the most reliable men and boats to carry out joint salvage operations, provided that they agree to terms similar to those which I have concluded with these salvors'. Nor might this involve the country in extra expense. Robbé was honest enough to admit that this seemed to him to be the best solution, because the 'employees' from Terschelling had proved unable to cope with such complications, as they were too few in number and the wreck was lying in an out-of-the-way place.

Though this difficulty was solved, another caused the Receiver of Wrecks of Terschelling a great deal of trouble: the Committee was slow in paying out the salvage money, so that the salvors complained. Robbé feared – and certainly not without justification – that this would affect their honesty, which already, as he humorously wrote to the Committee, 'has more than enough room between here and IJzergat'. However, he advised in the strongest terms against payment in the form of salvaged coin, 'for this is as black as pitch' and 'not being common currency' could not be brought into circulation, especially not before everything had been recovered.

In the meantime the salvage operations were not proceeding very smoothly, and when the returns diminished the complaints piled up. In the end, Robbé had to request that the salvage money might be sent without delay, 'as both the boat-owners and the common citizens are trampling my door down, as it were, with entreaties (that I will advance payment). If it is not forthcoming, a very violent storm will be blowing up for me with these folk, who are otherwise going about their work in a perfectly harmonious and orderly manner.' Yes, the salvage work was being so well organized by this time that no abuse was made, a state of affairs which the Receiver of Wrecks would certainly be able to maintain if he were 'continually' in a position to 'meet his liabilities to them'. One has the impression that the work of recovering valuables certainly was being carried out in an orderly manner. We see that the salvaged gold and silver bars were sent regularly to the state mintmaster at Dordrecht and Suttorp found a receipt dated 15 September 1800 in which this official confirmed the delivery of 29,248 Spanish piastres, 13 gold bars and 15 silver bars, worth 216,870 guilders. A receipt dated 3 December 1801 shows a total amount of 301,721 guilders. The first amount – as Robbé wrote – would have come from the agreeable catch of 'a nice bit of capital' on 4–5 September, two successful days. Then twelve gold bars, among other things, were hauled up out of two small casks, one of which came to the surface with its bottom stove in. On this occasion, the Receiver of Wrecks spoke very respectfully about the work of the salvors, which he called 'so hard and dangerous that, without having seen it with one's own eyes, one could never conceive what it is like'.

The difficulties of the salvage work were largely due to the fact that the packing around the bars and coins proved to be unequal to a protracted stay under the waves, which had already harassed the wreck for nearly a year. Robbé complained about the flimsy casks bound with wooden hoops which were so very fragile that 'only the other day some loose ditto hoops, staves, loose money and several loose bars of silver came up'. As for the casks in which the gold was packed, these were not very strongly made either, and the iron hoops proved to be so rusty that they fell apart at the least touch.

These handicaps weighed more heavily when the results of the salvage operations plainly diminished. In 1801 Robbé spoke openly of 'misfortune' in his letters, in which hope alternated with disappointment. Sometimes hope flared up for a little while, as on the day when one of the fishermen 'on hauling up a length of cable, found that he also had in the pincers a piece of paper which was folded flat and tied up, in which were several small items of Spanish gold coins'. Unfortunately, their expectations were disappointed and he went on dejectedly, 'Nothing has done more to discourage the fishermen so far this year than that – as happened last year – favourable tides and clear water do not coincide with favourable weather.' Although the wreck proved to be still partially intact half a century later, in 1801 the sea had already damaged it to such an extent that the contents of the *Lutine* had become scattered around and about the wreck. The fishermen had to try to grope for objects blindly and at random at a depth of about twenty-five feet, and it became obviously desirable to engage divers, for whom there was now more room and less danger, 'through the continuous disposal of rotten rigging, rusty guns and sabres'. According to Robbé, the only impediment consisted of 'the great number of stumps of timbers belonging to the wreck which jut out of the sand here and there and which one cannot see when the water is thick, but which one can almost feel and so avoid'.

But various objects were still being salvaged in 1801. Between 20 and 30 August 1800, Hendrik Blom (also called Bloem) and Co. had been doing salvage work, whereupon 'the citizens Keizer and A. A. de Vries Robbé' set off for The Hague as early as 4 September with the proceeds 'both of the small and of the large company'. Between 4 and 19 September fresh recoveries were made and these finds were taken to The Hague by Johannes Brandenburg, town clerk of Terschelling.

After this no more fishing seems to have been done that year. Between 11 and 14 July 1801 fishermen from Terschelling and Urk, acting 'in compagnie', recovered various items, which were fished up by a certain J. van Keulen.

That was the last of the fortune for some time to come. We hear nothing about the year 1802 and in 1803 the *Lutine* showed herself for the first time to be unpredictable and unresponsive to those who sued for her favours. In a letter which Robbé wrote on 28 May of that year, he declared, according to Suttorp, that it was now really essential for a good diver to be engaged, since 'all the casks which contained metal have fallen to pieces because of the passage of time and because of continual working on them, so from time to time bits of rusty iron hoops, a single cask of ingots, and things of that sort are hauled up, but no metal can be grasped with the tongs or irons, although, as one fancies, it can be felt just underneath the sand'. From the Receiver of Wrecks' letters we gain an impression of how the fishermen reacted, how their dejection alternated with hope.

'At this present time it is disheartening work for the salvors; in the past eighteen months they have been working on the wreck hundreds of times in vain, a great deal of equipment has been lost, too, and many heavy buoys and seamarks, some even having heavy iron chains on them, have been lost as well, besides the other expenses. However, the repeated encouragement that a lucky hour or two can make up for everything keeps them going.'

But those lucky hours were slow in coming. The storms which raged over the North Sea islands in November 1803 brought the sea over the shallows into such a turmoil that the situation on the wreck changed considerably to the salvors' disadvantage. An investigation showed 'that the part of the wreck in which one is accustomed to find the precious metals had now been covered by a large piece of the side of the ship (which had previously been found hanging more or less at an angle), thus impeding the salvage work, which was otherwise possible'. Attempts had been made over and over again to lift the piece of wreckage, Robbé explained, but the great depth of water in that place and the weight of the already saturated wood (which might also have become stuck fast somewhere) made all attempts fruitless. 'This business has caused the loss of a great deal of heavy equipment, all for nothing.' Apparently work then came to a standstill. Besides – as the reports show – the wreck became more deeply buried under the sand.

If one takes stock of the results of the first salvage attempts, the work was really only successful in the year 1800. It is all the more surprising, therefore, to read that the accounts for that year still showed a deficit of about 3,241 guilders. Of course, the salvors' wages (which *were* paid) formed a very important item, but in the *Short but Itemized Statement of Income and Expenditure* which F. P. Robbé, in his capacity as steward of the State Properties in Terschelling, caused to be sent to the Committee, in compliance with instructions dated 4 December, there were other and more important items of expenditure. Reference is made to the 'repeated taking of soundings and the placing of seamarks on the wreck of the *Lutine* last autumn', and also 'this spring after the ice-drft'. Expenses were incurred 'on account of several advertisements on Vlieland, Ameland, here and elsewhere' in order to check 'unlicensed fishing' on the wreck.

That the salvors made use of landmarks to find the place where the sunken *Lutine* was lying is proved by the payment made 'to the beacon-keepers for setting up the beacons during this past year, whenever the boats went out to fish over the wreck'. And that Robbé also took the necessary safety precautions is shown by the payment made to 'the signallers for keeping a constant watch and reporting when the ships were over the wreck of the *Lutine* and returned from the same'. Indeed, these functionaries literally kept a weather-eye open even at night: 'They keep watch at night by clear moonlight and make a report should they have observed or suspected that strange boats were making for the wreck.' That they only made a check on moonlight nights is explained by the fact that not a single fisherman stood a chance of finding the place of the wreck in complete darkness. No landmarks could then be discerned, supposing that the beacon-keepers had neglected to take away the *Lutine* beacons (on the beach or among the sand-dunes) in time.

After accounts showing 'poor-rates carried forward and also both private and other disbursements for keeping watch, correspondence, assurance of scrupulous delivery of the abovementioned metals and more of the same nature over the entire year' – all of which shows again how the salvage operations were organized – we finally come across an item which can be directly related to the tales about wagonloads of gold and silver and the wholesale polishing of blackened coins, which the old inhabitants of Terschelling told to Acket about the year 1880. For Robbé also accounts for the payment of sums of money 'to the warehousemen, porters and workmen for cleaning and arranging the metals and coins, also transport vice-versa', added to which were the harbour dues.

Although, as Dr Suttorp said, the poor results and high costs brought the salvage operations to a halt before long, it should be remembered that, leaving out of account the salvors working at their own risk, a good many men of Terschelling, acting as assistants, porters, workmen and carriers, profited by the short-lived boom which the island witnessed after the summer of the year 1800. That an impoverished population should have kept fresh the memory of this happy period is not surprising.

11 Pierre Eschauzier fishes for a fortune

For a long period after the successful salvage operations under Robbé there is a lull. Only when the period of French domination has come to an end does a new promoter come on to the scene. He is Robbé's successor, Pierre Eschauzier,[26] Chief Receiver of Wrecks of Terschelling. As early as 1814, a few years after his appointment in 1810, this official – well-born and evidently a man with influential connections – addressed himself to King William I, to whom he unfolded his plans. In a decree of 16 June the sovereign, 'on the recommendation of the President of the Councils and Auditors-General of the Crown Lands dated 6 June 1814', makes known that he 'authorizes the Chief Receiver of Wrecks of Terschelling to make a further attempt at salvaging the cargo of the frigate *Lutine* in connection with which some investigations have already been made, and allocates the sum of 300 guilders for that purpose, with a recommendation at the same time to the above-mentioned Receiver of Wrecks to use all possible caution in the choice of persons employed in this task'.

It is reported that the Chief Receiver of Wrecks 'made twelve successive attempts with various boats in order to find out whether the place where the money lies was uncovered'.

But there the matter rested that autumn, for the account which Eschauzier (this time referred to as 'provisional' Chief Receiver of Wrecks) sent to the finance department of the Crown Lands on 2 March 1815 reports only '8 Louis d'or and 7 Spanish piastres fished out of the wreck of the *Lutine*', whereupon the auditors hurriedly noted 'The person rendering this account should be directed to send this specie in kind to the board.' Further information about success or the lack of it has not come our way. Thus there are no grounds for Martin's assertion in his *History of Lloyds* that Eschauzier devoted seven whole years to salvage operations, only fishing up 17 coins in all. This number is the exact amount of the booty which had already been hauled up in October 1814, while the hypothetical seven years' activity must, it seems, be linked with the fact that it was only in 1821 that new plans were made.

12 A royal decree and an association

It is not known why apparently no salvage attempts were made on the *Lutine* in the first years after 1814. Perhaps the wreck was lying deeply buried under the sand again. Perhaps, too, the Chief Receiver of Wrecks, Pierre Eschauzier, realized that hooks, pincers and scoops would be unable to haul up anything of importance. At any rate, the summer of 1821 had arrived before he unfolded new plans. This is apparent from a lengthy letter which Eschauzier sent to the governor of North Holland on 25 June of that year. Reading this document, one forms the impression that there were two motives deciding the writer to make a new attempt: fear of competition and the fact that in England a diving bell had been used with success. The letter seeks to make plain in the first place that the wreck lies 'in the territory of which I am Chief Receiver of Wrecks' and that he, Eschauzier, still has the right to salvage there 'to the exclusion of all others'. Apparently this reminder was necessary, for rivals seem to have been in the offing. One of his arguments was that 'at no time since the assignment of this fishery has the Chief Receiver of Wrecks of Vlieland offered any opposition', not even at the time when several hundreds of thousands of guilders worth of gold were recovered in two successive years (that is, in 1800–01). Had his colleague in Vlieland recently been showing an undesirable interest, then?

More important is the announcement that Eschauzier had drafted 'an effective plan' for a new salvage attempt, 'i.e. by using a Machine'. He had come across a description of this apparatus in the *Vaderlandsche Letteroefeningen* (National Letters) of July 1820 (no. 10)[27], and he now gave the King's representative to understand that he would 'procure' it 'with the required competent amphibious ["amphibicque"] Englishmen who make this underwater journey daily, together with such persons as will be needed to control the pressure pump and the crane'. He had already opened a subscription for 10,000 guilders for its purchase and expenses and wrote that he had no doubt he would reach his goal. All guarantees were present for an under-

photo: *Musées de la Marine, Paris*

Model of the French frigate *La Dédaigneuse*, sister ship to the *Lutine*, built in 1766 and armed with 26 cannon.

Engraving of a ship with Taurel's Dutch diver (Brand Eschauzier, 'Communication', 1859).

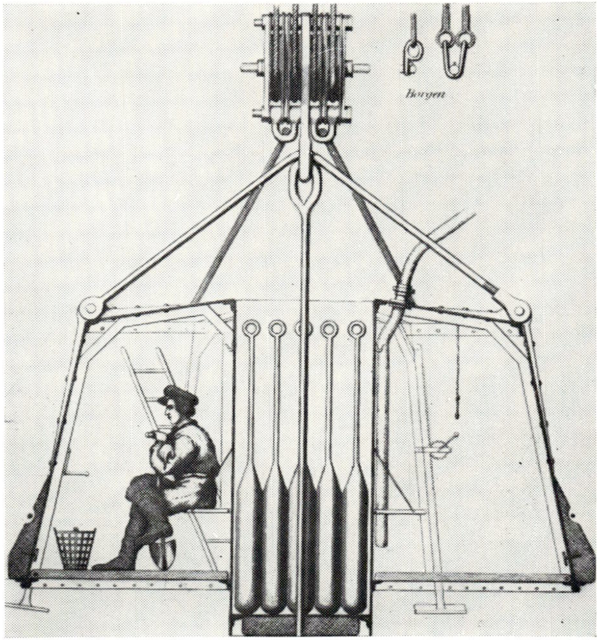

Engraving of the diving bell designed by Taurel (Brand Eschauzier, 'Further Communication', 1861).

Engraving of Ter Meulen's sand-diving process (1882).

The mill on West Terschelling with the *Lutine* landmark which was washed away in 1863.

Brandaris lighthouse with the landmark erected in 1874 on the Noordsvaarder sandbank.

The Veldkaap on Vlieland with the mark erected in 1874 in the foreground. (Drawings by Ter Meulen, 1874.)

taking from which the state would also profit, because it 'will receive one out of every two guilders'.

The governor sent the letter on to the King and in the meantime Eschauzier went on organizing the enterprise. In July 1821 he wrote very elatedly to his 'Très Chers Frères' that there were already nineteen participants who had promised sums ranging from 100 to 600 guilders. Some of the subscribers met with Eschauzier at the home of Sjoerd S. Wijma, notary, in Harlingen, on 16 July 1821. They had now cast their eyes on a new method employing a diving bell. A sum of about 10,000 guilders was required, for which shares of 100, 200 and 300 guilders and over would be issued, and the founding of an association necessary, with P. Eschauzier as director. All precautionary measures would be taken to guard against fraud: the salvaging might only take place in the presence of the board, consisting of the parties present; immediately after being salvaged, the gold and silver bars and the specie would be deposited in chests 'with two different approved locks', one key to which would be in the keeping of the parties present (constituting the aforesaid board) and the other in that of the Chief Receiver of Wrecks. Once the chests had been brought ashore, they would be opened in the presence of the parties present, who then 'will make a careful inventory of the contents in the presence of a notary'. This agreement was registered the very next day and they set to work very quickly. Their first task was to locate the wreck once more. On 5 August, 'before Terschelling and the Vlie near the *Lutine* wreck', after having fished above and near the wreck on various occasions, Zeijlemaker, Reedeker and a certain C. Jongebroer (a surety) were able to report triumphantly that on that day they had had 'the most positive traces of the same in their hands', to wit, two pieces of iron ballast, a cannonball and lumps of rust. In addition they had 'clearly seen the whole extent of the wreck and also of the anchors which have been lying before it for twenty-one years, in the same direction as formerly, and that in twenty-six and twenty-eight feet of water'. Although in the meantime the weather had become more settled, the sea even having been very calm on the two preceding days, they had not yet succeeded in boring through 'the hard layer of rust' which was everywhere as the pincers and tongs kept losing their grip. For the time being they were still trying in the old way.

While the board was thus occupied, the Minister of Home Affairs was investigating the matter in consequence of Eschauzier's letter to the governor of the province under whose jurisdiction Terschelling at that time fell. This investigation evidently led to a favourable decision, for on the basis of the Minister's report, dated 10 September, King William I issued a royal decree, which must certainly have come as an agreeable surprise to Eschauzier and his partners. For the royal decree conferred an important privilege, indeed, no less than a monopoly, which authorized Terschelling's Chief Receiver of Wrecks, in accordance with the plan laid before the governor in his letter of 25 June, 'to attempt the further salvage of the cargo of the English frigate, the *Lutine*, which foundered between Terschelling and Vlieland in the year 1799, proceeding from London and bound for Hamburg, and having a very considerable capital on board, consisting of gold and silver coins, believed to amount in all to 20 million Dutch guilders'.

Conditions were naturally stipulated. The state must not incur any expense; half of what was brought to the surface would be allocated to the enterprise and the other half to the state, and the chests brought ashore had 'to be inventoried in the presence of a notary with the utmost promptitude'. Furthermore, the beneficiary was required to make a return of the articles salvaged to the governor of North Holland immediately after every salvage attempt. Finally – this, however, is not in the royal decree but is taken from the minutes of the gathering of the Royal Netherlands Institute of Engineers on 12 June 1856 – Captain H. J. Ortt, superintendent of pilotage north of the River Maas, was instructed to keep watch on the government's behalf on the operations with the diving bell and thereby protect the interests of the state.

In the meantime, the diving bell had not put in an appearance. From the statement of accounts of the enterprise between the years 1821–30, preserved in the municipal archives of Terschelling, it appears that Eschauzier paid a visit to Brussels in July 1821 (then the seat of government), to Haarlem (for an interview with the King's governor) and to Amsterdam, where he lodged for six weeks while organizing the new enterprise. In September he set off for London, for which he also declared six weeks' travelling and lodging expenses, justifying this prolonged stay by pointing to 'the decease of Mr Rennie, the engineer'. When Eschauzier set off on the return journey to Terschelling in November 1821 he had apparently succeeded in talking to the British engineer before his death. This appears from a long letter dispatched on 22 January of the following year to the governor of North Holland – who seems to have been well-disposed towards him –

in which he tried to obtain government support for the use of the British diving bell. He had found out that Rennie, who until then had been using catheads (beams projecting over both bows of the ship with sheave-holes on which the anchors were hung when raised to the hawse-holes), had in the meantime changed to another method. He now suspended the bell, as he had shown his Dutch guest in a drawing, from the jib-boom of a schooner, which had already been used in Plymouth for removing reefs at a great depth in the sea and for making a dam thirty-five to forty feet under water. Eschauzier had now heard from some Dutch naval officers familiar with this 'machine' that the schooner *Gelder* was at present lying in the Nieuwe Deip 'without any employment', and this vessel could be extremely useful for the salvage operations. He therefore requested the governor to propose to the government that they should put the *Gelder* at their disposal for the duration of the undertaking, and allow them to send the vessel to London, where Rennie's associates would then be able to adapt it for salvage work. To make the proposition more attractive, the petitioner added that Lord Cochrane was planning to use a similar machine to salvage valuables from a man-of-war which had struck a reef off the coast of America and sunk in deep water.

The *Gelder* was not placed at the enterprise's disposal, and on 11 April they purchased the schooner *Minerva*, renamed *Duiker* (Diver). A printed report to the shareholders dated 20 November 1822 contains the information that the schooner arrived at Terschelling with the diving bell towards the end of June. Everything had been found to be in order, 'all the equipment was pre-eminently satisfactory and was indeed excellent'. Yet they had no success. While the wreck had lain bare in the months of March, April and May, it was feared that 'the necessary depth would not be maintained in the summer'. The facts would soon prove them to be right, as may be gathered from the circular of November 1822. In it we read:

> 'Until 19 July there was no opportunity open to the fishery, because of contrary winds and unsettled weather. On that day conditions were favourable and the bell was under water for three hours at a depth of twenty-three to twenty-four feet. Mr Cruso, the manager, had already discovered the anchors of the wreck and recognized them as belonging to the British royal colours, when a gathering thunderstorm forced Captain Kerkhoven to have the bell hauled up in order to reach harbour, which they were barely able to

make in time. From that day forward, in spite of all one could wish, nothing could be done for three weeks because sand was piling up over the wreck and winds were contrary. About the middle of August an attempt was made with the schooner on another wreck, off Texel, which, however, on closer investigation proved to be hampered by shoals. On the report that there was an extra fathom of water over the *Lutine* again, she returned to Terschelling; but when fresh soundings were taken there was renewed disappointment and it was obvious from this that if there should be a sufficient depth of water on a single day in summer, it is very quickly shoaled up again.

'In the meantime the schooner had had to have some repairs carried out. At the end of the month of October the wind and weather were favourable, but on making further investigations it once more turned out that the tempestuous seas and southerly winds had piled so much sand over the wreck that nothing further could be hoped for this season.'

After this melancholy account Messrs Eschauzier, Zeijlemaker, van Kammen and Reedeker found it necessary to put some heart into the other shareholders: it would not be feasible to put a stop to the enterprise now under way, especially as the situation at the wreck would certainly be more favourable in the coming spring. For then the north-westerly winds of the winter might well have 'blown away' a ridge of sand 'which settles down on the wreck with south-easterly winds'. Besides, it was known with some certainty, on the basis of the insurances taken out in London in 1799, that in spite of the prior salvaging of 'eight hundred thousand guilders' (an over-estimate, as we have seen), a very considerable amount of treasure must still be present in the wreck. Under this layer of hopeful gilt the shareholders then found the somewhat bitter pill: for the time being, a further 15,400 guilders would be needed for loan redemption and for making another attempt the following spring, and also for repairs to the schooner. Or, in the tactful words of the board: 'one would be able to make do for the time being' with this sum. Those who did not want to put more money into the undertaking would, of course, receive proportionately less from the possible share-out. In future, money could be deposited with Messrs Ueberfeld and Serrurier, who would act as correspondents for the enterprise in Amsterdam.

The invitation met with results and regular salvage attempts continued. This is shown by the records which give yearly particulars. Skipper Pieter Visser from Terschelling appears in the books from August 1822 to the summer of 1829. This man was the owner of the lighter or ketch *De Vrouwe Anna* (Lady Anne) and signed a contract with the enterprise 'for the carrying of a diving bell'. Messrs Eschauzier and Co. had to insure his ship (worth 9,000 guilders) against all damage and pay Visser six guilders a day, and if he was successful, ten. *De Vrouwe Anna* seems to have replaced the schooner.

The divers, Cruso, and Captain Kerkhoven were kept on after the 1822 season so that their work could be continued in 1823. It is surprising to learn in this connection that in addition to the salvage work being carried out by the Cruso group, relations were also entered into with other Britons, who did not work with a bell, however, but with tongs and scoops. A note among the documents confirms the arrival on Terschelling in June 1823 of a certain P. Ellis. In the course of an interview at the home of P. Eschauzier, in the presence of Reedeker and Kerkhoven, the Briton first demanded 25 per cent, then 20 per cent of what was salvaged. Ellis was accompanied by two British fishermen and they planned to strip the wreck of sand. A large number of fishing boats with crews of 400 men on board were to come to Terschelling for the purpose. But for the time being not more than 30 men turned up, having pincers, scoops and poles with sacks on the end. At that time a mere three feet of sand was lying on the wreck. The note concludes with the text of an agreement which was signed on 29 July 1823 by J. A. van Zuylen van Nyevelt (court registrar of Haarlem) for the board on Terschelling and by Peter Ellis as proxy for the British fishers, Wm. Whorlow, the four Bell brothers, and W. Hoult of Whitstable.

This work proved completely fruitless, in both 1823 and 1824. In 1824 a third diver was engaged and we hear nothing more about the fishermen from Whitstable and their traditional salvage equipment. Work also went on in 1825. In a circular sent out from Terschelling on 15 April 1825 the board reports, among other things, that skipper Visser had taken regular soundings in the course of the previous summer, but that, nevertheless, 'all the activities of the last-mentioned year have been fruitless, because of the natural hindrance offered by the shallowness of the water over the wreck, which, though varying a great deal, caused all reasonable hope of success to fade time and time again'.

The board would have considered the suspension of activities for the time being, had it not appeared on 31 January 1825 that the depth on the cross-bearings had increased. The sandbank had also 'completely shifted and withdrawn itself from the same', so that the members of the board had fresh hopes that 'the wreck would perhaps soon be found to be uncovered'. However, more cash was needed – in fact, 20 per cent of the capital available at that moment. That money seems to have been forthcoming, too, for twenty-two 'sea trips' by fishing boats from Urk are accounted for in 1826, as well as outlays for crew for the crane, machinery, wages of divers, etc. But in the summer of that year work seems to have been broken off for reasons unknown to us, for in July 1826 an item is entered connected with the fact that Visser had agreed 'to the hire and custody of the bell and machinery in complete working order until 1 March 1827'. He also received 1,000 guilders for 'the hire of the lighter until 26 November 1826 in accordance with contract'.

However, in 1827 they set to work again with fresh courage. From the beginning of March until 17 August, trips to the *Lutine* and the 'iron ship' (apparently another wreck in the vicinity, which may have given its name to the IJzergat) are noted in the account book, of which two nocturnal trips, entered as such, receive special attention, while two divers appeared on the wage-sheet for 'descents with the bell'. Young Zeijlemaker wrote up the log-book (unfortunately not preserved). Whereas all those who were actually involved in the salvage operations were paid for their labours, the members of the Terschelling board lent their assistance for nothing. This being so, no one will have taken it ill of the gentlemen for bringing the following point to the attention of the shareholders in rendering their statement of account: 'The members of the board would be justified in submitting an annual account for 600 guilders for a great many journeys, both to sea and to Amsterdam, for loss of time, fatigues and dangers endured, etc., and thus charge 3,600 guilders over a period of six years, but seeing that the enterprise has yielded no profit to date, they therefore waive this claim. Ergo nihil.'

We have not been able to find any particulars concerning 1828, but in the following year the salvage attempt appears to have finally come to an end. On 13 April 1829 the sum of 4,200 guilders was paid to skipper Visser 'in final settlement for the hire of the diving ship', this sum being obtained 'from the proceeds of the lawfully sold bell and appurtenances'. With this, the accounts book for this period closes. From a letter from Eschauzier, dated 1835, it appears that the diving bell had been sold to the Dutch Navy.

We cannot end this chapter without mentioning the reaction which the setting-up of the enterprise, the granting of the concession to Pierre Eschauzier and the commencement of the salvage operations aroused in England. A number of influential underwriters from Lloyds began to busy themselves with the affair. They argued – not without justification – that if a good part of the bullion were still in the wreck, it belonged to them. They were also of the opinion that the Dutch government might now reserve the cargo to itself. The Committee of Lloyds finally acted and approached the British government on this score. The latter opened talks with Brussels, in the course of which the Netherlands government was requested to make over half the valuables to be salvaged to those on the British side who had a claim on the cargo. Mr William Bell, the chairman of Lloyds Committee, was extremely active in this affair and even made a journey to the Netherlands during which he visited the scene of the wreck. Bell experienced the pleasure of learning from a letter which Mr F. Conyngham, Secretary at the Foreign Office, wrote to him on 6 May 1823 that the King of the Netherlands was prepared to make over his half share to the English. The British underwriters who had a claim were further advised to come to an arrangement with the Eschauzier association. In addition, it seemed best to Mr Conyngham that his fellow countrymen should fall in with the King's offer.[28] Once the British underwriters had declared themselves willing, the matter could be settled.

On 23 May of that year William I of the Netherlands issued a decree revising that of 14 September 1821. Regarding the preliminaries, the document states that the British ambassador had sent a memorandum in which permission was requested on behalf of the English government 'to be allowed to fish for the cannon, together with the specie and bars which were on board the frigate mentioned'. The substance of the new decree was that everything that in accordance with the royal decree of 14 September 1821 'had been reserved to the state from the cargo of the above-mentioned frigate' was ceded to the King of Great Britain, but only as a token 'of our friendly sentiments towards the Kingdom of Great Britain, and by no means out of a conviction of England's right to any part of the aforementioned cargo'. A report by the Dutch Minister for Foreign Affairs was also cited, from which it appeared that the English monarch intended to make over the share bestowed on him to 'the Society of Lloyds, which will send a proxy to reach an agreement with the Dutch parties concerned with the "fishing-up" of the above-mentioned cargo on the putting into effect of the decisions arrived at'. This, in fact, is what was done, so that ever since 1823 Lloyds of London have had a finger in the pie at every attempt to salvage the *Lutine* treasure, without, however, ever having shown much activity themselves.

13 Britons above the wreck

If we are to believe N. S. Binnendijk, who in December 1935 wrote a few articles in *The Log*, a Lloyds magazine, this company, prompted by the failure of the Eschauzier venture, sent a number of British divers to Terschelling. They investigated the situation but made no attempt at salvage. Martin also mentions this move in his *History of Lloyds*, although we have been able to find no other information about it anywhere. But there was a British intermezzo in the period 1834-36, and both a circular issued by the board of the *Lutine* enterprise and the second balance book, covering 1834 and 1835, yield information about this. In May 1834 two pilot barges made three journeys 'to pinpoint the wreck' and on a fourth trip it was discovered that the *Lutine* was lying clear. Hereupon the Board immediately invited the Bell brothers of Whitstable 'to come hither with their apparatuses' and resume operations. The Englishmen accordingly arrived on Terschelling in June of that year accompanied by two 'shallops, pinnaces, and their crews'. On board with them they had a diver, with a newly invented 'machine'.

This diver descended to the wreck a number of times to no avail, making it evident to the members of the Terschelling board, who followed his every movement with critical eyes, 'that his skill could promise scant success in this place'. And, indeed, after having made a dozen trips out to the wreck, the Bells' vessels returned to England, as the board reported, adding that the diving machine, which at times had stayed under water for more than two hours, would have had more chance of success had the men who went down in it not refused 'out of obstinacy' to enter the wreck, to 'break it up' as far as possible, or to clear it of sand. What is more, they had wasted time by going to work on the wreck of a Portuguese ship lying near by, which enterprise had produced no more than one plank and a length of rope.

When the spring of 1835 had come the water above the wreck proved appreciably deeper, while repeated investigation with the aid of pilot boats and fishing vessels had established that the sandbank where the wreck lay – and in the area surrounding it as well – had shrunk considerably in size and shifted southwards. Furthermore, hooks and poles had produced clear evidence that part of the wreck lay clear of sand. The absence of the English diving bell was then 'lamented', but no further word was received from the Bell brothers. Instead an offer was received from a certain John Bethell, of Mecklenburg Square, London, to undertake the salvage 'with two even improved divers' air-machines and two qualified workmen' on what, moreover, were much more favourable terms. In the records we find a copy of a letter, no doubt from P. Eschauzier, and dated 9 April 1835, in which he reports to Bethell that the day before he had gone to the site of the wreck with two pilot boats. Accompanying him were a Mr Hofmeester and O. J. A. Eschauzier, lieutenants in the Dutch Navy; G. S. Rotgans, a captain in the mercantile marine and S. Pottenga, quarantine officer. The vessels had become stuck, as the report has it, at twenty-seven feet at the low tide water level, i.e. thirty-four feet at high water. The swell, however, was too great to allow them to make further investigations with thirty-six foot tongs, so a heavy stone was laid on the wreck and a 'watcher buoy' left.

In a letter of 9 June to the Englishman, it was stated that there was now thirty feet of water at low tide. Bethell must have sent his men shortly after this. The accounts for July show various entries relating to this expedition. The sum of 90 guilders and 35 cents was paid as import duty on the 'machine' and 3 guilders to the divers for 'inspection, manipulation and clothing'. Mr Norton, referred to as the head diver, received 48 guilders for four weeks, while there were freight charges for 'fetching and paying for machinery from Amsterdam'. A trial descent seems to have been made according to an item reading: 'To the English divers, as an encouragement for the test before the harbour: f.3.00.' Two fishermen from Urk made three journeys to the wreck and the skipper, Jan Doeksen, received the price of the hire of a ketch for 42 days (f.3.00 a day). Norton's salary was still paid for September, after which the balance book records nothing. As to the results, the board was obliged to report to the shareholders that: 'The two Englishmen proved afraid to go down and however calm the sea was and however favourable the opportunity, nothing could reassure them ... They declared that they had only practised once or twice in inland waterways and that they did not dare to descend to the bottom of the sea in this spot.'

The British director promptly sent the men home, dispatching an urgent request for two experienced sea divers, 'which he said Mr Bethell did indeed have in his employment'. The Chief Receiver of Wrecks added his word too, and the board wrote that it was expecting the new divers any day, 'while the English director is staying at Terschelling with his diving-machine in order to re-commence work immediately they arrive'. This was written on 12 September and the board was still hoping to use 'the favourable opportunity in the month of October', when the weather was sometimes fine and calm, while they anticipated new operations lasting three months in 1836. But on 28 September 1835 P. Eschauzier had to inform Serrurier that 'for the time being Mr Bethell had no divers available', although his agent, Captain Norton (address: Lloyds Coffee House, London) had committed himself to return in the spring with a diver and a machine 'perfectly suitable for the coast'. Norton had departed for London the day before but had left the equipment behind on the island. As regards the new divers, they would have to be housed and fed on a boat hired by Norton 'to deprive them of the opportunity of rendering themselves unfit to carry out their work owing to abuse of gin'! Although since 1814 Eschauzier had spent much time, money and energy on the *Lutine* affair without any result, his optimism still remained unimpaired, witness the final sentence of his letter to the banker: 'So we have never had more splendid prospects than those for the spring, when we shall go to work with Strength and Energy.'

No wonder Eschauzier was keeping London interested. On 4 October he wrote to Norton that the diving bell that Rennie had made, and which had been sold to the navy, was again at the contractors' disposal, while he also mentioned a new plan in which a famous engineer was to have a hand. On 27 October the Chief Receiver of Wrecks also approached Bethell's representative once again. The latter apparently failed to reply, so in March 1836 Eschauzier wrote to Bethell direct, asking why he had not responded to his letters of the previous October. It was only much later that it became clear why the Englishman had not replied. A circular issued by the board on 30 July 1843 merely mentions in a single sentence 'that the work was brought to a halt owing to the absence of Mr John Bethell's divers'.

The Chief Receiver of Wrecks of Terschelling, the Noordsvaarder and Griend was spared further disappointments, for on 21 June 1837 he died, and was buried in the churchyard at West Terschelling, in the shadow of the Brandaris lighthouse. A monument bearing his name and the offices he had held was erected on the family grave.

Although Pierre Eschauzier's name is inseparable from the history of the *Lutine* we know little of his personality. His letters and circulars demonstrate the enthusiasm with which he devoted himself to the enterprise, which had been begun on his initiative. Was he prompted by greed alone? Was it true that he had had himself appointed on Terschelling in order to be able to make money out of the *Lutine* wreck? Instead of attempting to answer these questions, it would, we think, be more to the point to quote the verdict Martin gives in his *History of Lloyds* – even though not innocent of a British sense of superiority: 'a man not only enthusiastic, but of enlightened views, and remarkably free from the national prejudices of his countrymen'. In his account Martin may have had Eschauzier die a few years too soon; nevertheless, his character sketch would not have been far out.

14 The men of Haarlem get a chance

The circular of 12 September 1835 received by the shareholders in the *Lutine* enterprise was to be the last they were to get for the time being from the board on Terschelling. Almost eight years went by before there was further cause for acquainting them with the position. Yet the board had not been idle in the interval. The famous bullion ship still lay at the spot marked, under a now deep, now not so deep, layer of sand, and this knowledge alone caused interest to flare up again.

In a letter of 17 September 1839 Wijma suggested to Reedeker the possibility of using the King's influence – the monarch was personally interested, apparently – 'to get the National Corps of Divers put at our disposal next spring to work on the *Lutine*'. This proposal was well received, for Reedeker, Van Kammen and Wijma submitted a request to the government that same year 'to be allowed to have the Navy's corps of divers put at our disposal from 1 May to 1 October 1840'. The new Chief Receiver of Wrecks on Terschelling, Burgomaster A. Swaan, had rather high hopes of this. In a letter of 1 November 1839 he openly admitted that: 'I only wish the Corps of Divers was already at work and the rusty or deeply sunk gold or silver bars might see the light of day, we could then make right merry.'

The decisions of the Ministers of Home Affairs and of the Navy, dated 8 and 18 May respectively, hardly gave much call for rejoicing. Both replies contained a refusal, as appears from the circular of 30 July 1843. This circular also makes it clear why they wanted to call in the navy: less than a quarter of the additional contribution asked for had come in, so the enterprise was not in a position to launch further attempts itself. In addition, the board, which after the death of the founder, Pierre Eschauzier, had preserved 'the co-operation with the heirs of the said Mr Eschauzier regarding the rights in the enterprise, as they trust', had done everything possible to achieve their aim without involving further sacrifice, but were still 'without any efficacious means through lack of funds'.

So it would have been with relief that they learned of the offer made in spring, 1843 by the firm of Van Geuns of Haarlem, manufacturers of gutta percha and owners of a sea diving apparatus, to attempt to raise the treasure on conditions which were eagerly accepted. The board's report makes no mention of the danger of competition which was threatening at this time. For in October 1842 Th. Denny Sargent, a civil engineer in Boston, had petitioned the Dutch government for exclusive rights to salvage work on the *Lutine*, offering to pay one fifth of the proceeds into the national exchequer. Since the *Lutine* affair had almost been forgotten in The Hague, this request resulted in a thorough enquiry. But for reasons unknown the American made no further approach.

In the meantime Van Geuns had made a beginning with their experiment, proceeding in the traditional manner 'with pikes and tongs, and had, indeed, pulled up a piece of the inner lining of a gun-port, attached to which were a few sheets of rust and countless cannon balls rusted in their turn on to these ... and also discovered that it was not lying too deep in the sand'.[29]

As early as 30 July the board was obliged to report to the shareholders that: 'To begin with, the attempts to trace the wreck were successful; the same was found to be far more clear of the sand and they had the satisfaction of bringing a number of objects (sufficiently publicized in the press) to the surface; since when the prolonged bad weather and the northerly and westerly winds blowing since the first of June until the date of this communication have prevented the diver from making any further attempt.' This had also exhausted the funds collected for the purpose, so that there was another request for money in order that the new contract with the Haarlem manufacturers should be adhered to in the months from August to October 1843, and from March to May 1844. The board tried to tempt the shareholders by reminding them that 'conditions at the site of the wreck had never been so favourable since 1823'.

The effect of this circular was apparently disappointing, for we read nothing further about the Van Geuns project until 1845. In a letter dated 6 October of that year and addressed to Serrurier, the banker, the notary Wijma reports that Van Geuns had offered his services once more 'to set to work again with improved apparatuses'. This would cost the enterprise nothing, but in this case the Haarlem manufacturer wished to receive more remuneration, i.e. 30 per cent of the half share to be received by the board. The Harlingen notary thought this too much, and Lloyds were still claiming half, although they were taking no part in the fishing. Apart from that, he was optimistic about commissioning Van Geuns once more. After all, his operations on the koff *Geetruid*, which had foundered on the Noorderhaaks Bank, had met with success. The cargo of rails the vessel had on board had been largely salvaged by Van Geuns' divers. Serrurier, too, in his reply of 16 October, wrote that Van Geuns and his men 'would be able to lay their hands on the wreck's treasure' but that Van Geuns himself did not, alas, wish to incur any financial risk while 'experience has shown that it is difficult to make him any advances and he finds it impossible to risk anything'.

When the *Lutine* enterprise sent no further word about the financial side of the matter, J. Van Geuns tried another way. In July 1847 he petitioned the King – albeit with the support of the board 'of shareholders in the *Lutine* enterprise' – to appoint an official committee of experts, to whom he could explain how he 'judged the salvage of the cargo to be possible and practical in view of the advances in the art of diving'. What measures would then have to be taken in the interests of the rightful claimants and the motherland in general could be left to the King to decide. The government, which were then, it seems, of the opinion that the enterprise's rights had lapsed as a result of the enactment of the Commerce Bill did not consider Van Geuns either to be qualified to attempt the salvage to the exclusion of all others, and since they had nothing at stake, his petition was refused. With this, the name of Van Geuns disappears for good from the *Lutine* annals.

The Commerce Bill also played a rôle in the case of the petition submitted to the Dutch Minister of the Navy in 1846 by two English divers, E. Hill and H. Downs, of London, who were also anxious to attempt the salvage. After a voluminous correspondence between three ministries and the governor of the province of North Holland, the decision arrived at, according to the *Handelsblad* of 27 April 1847, was that the wreck must be considered as lying in the Buitengronden (outer shallows), which meant that Article 550 of the Commerce Bill of 1838 applied – a mistake, by the way, which the government were to admit eleven years later. This meant that anyone was qualified to fish for treasure 'provided he handed over what he salvaged to the appropriate official in return for the reward fixed for that purpose'. It was decided to inform Hill and Downs that, moreover, half of everything recovered had to be surrendered to Lloyds. This was apparently such a disappointment to them that they did not repeat their request. After these applications by the two Englishmen and by Van Geuns a long silence settled over the *Lutine* question once again. It was not until 1857 that chance circumstances lent the matter a new topicality and led to success comparable with Robbé's at the beginning of the century.

15 The second half-million raised

In the spring of 1857 a buoy had come adrift in the Vlie Passage. When a Terschelling fisherman, Wever by name, hauled this buoy on board, he pulled a lump of wreckage with it which proved to be a piece of the forepart of the *Lutine*, and once again attention became focused on the bullion ship.

It seems that Wever and his colleague, Van Keulen, made it appear that it was they who had identified the piece of wreckage and so caused the sleeping dogs to stir. In a letter sent by S. Reedeker, G. S. Rotgans and W. S. L. Eschauzier to the Burgomaster of Terschelling and to J. P. Brand Eschauzier at Kampen on 16 January 1859 the writers declare that it is 'lies that Sjoerd van Keulen and Wever discovered the identity or furnished any clue to it'. The two fishermen had recovered a buoy to which a piece of wreckage was stuck. They took it to the harbour and 'when they were dismembering it and sharing it out among them', Reedeker 'recognized it as belonging to the *Lutine*'.

Having set matters right, the board admitted that the chance recovery of the buoy and piece of wreckage 'has done much to get matters moving again', and they accordingly agreed that both Lloyds and the enterprise should collect 240 guilders in order to pay this sum to Sjoerd van Keulen, Cornelis Wever and Gerrit Hakvoort. The last 'because he has on several occasions reported objects floating up from the wreck'. They may also have been persuaded to pay the three men a gratuity because only the year before descents to the *Lutine* had met with much success.

Two more figures are linked with the recovery of the second half million guilders from the wreck. The first was Jean Pierre Brand Eschauzier (Brand Eschauzier being his surname), born in Amsterdam in 1803 and at the time he was concerned with the salvage work already a retired captain of artillery and an ex-lecturer in French at the Royal Military Academy at Breda. He was then living at Kampen and died in The Hague in 1871. The second, resident in the same town, was the physicist (also referred to as a mechanical engineer) Louis J. M. Taurel.

They got in touch with each other during the summer of 1857 and this led, in July, to provisional soundings (on the 11th), the laying of a buoy (on the 15th) and a further investigation (on the 20th). This showed that a channel had formed straight across the Goudplaat sandbank, leading over the wreck, so that the latter was not merely clear of sand but had also sunk further below the surface with the channel. Secondly, the bows and stern, together with the decks and sides, had come completely away, leaving only the keel with the keelson above it and some ribs attached to this, which sections of the wreck had apparently remained in place due to the weight of the crusty layer of rust that was spread over them. A further general investigation took place on 5 August and this was apparently so promising that that same day, on Terschelling, Brand Eschauzier and Taurel proceeded to 'confirm in writing an agreement which had been entered upon several weeks before and was already being put into force'. Both parties promised each other mutual support and assistance in the salvage attempts and arrived at a financial agreement.

And so, on 13 August 1857, with a light SE wind, a start could be made. Taurel had assured himself of the co-operation of a group of divers from Egmond, who were headed by Teunis Panteydt, local director of the North and South Holland Lifesaving Society and owner of a firm that supplied victuals to ships. A man of action, Panteydt equipped a bluff-bowed fishing boat, a bom called *Egmonds Hoop* (Hope of Egmond), and sailed off in it to Terschelling with a diver on board. This diver was Aldert Wijker,[30] a pilot by profession, who was the skipper of the Egmond lifeboat as late as the eighteen-sixties.

Brand Eschauzier relates in his *Further Communication* of 1861 that the helmeted divers – given this name ('helmduiker' in Dutch) to distinguish them from bell divers (klokduikers) – to whose attention Taurel had drawn the presence of a lump of rust, obstinately refused to bring a few pieces of this rust to the surface 'due to a strange misconception'. They claimed that they were only *stones*. As a result, the first diving attempts proved fruitless, which caused Taurel, at his wits' end, to leave Terschelling on 16 August in the hope of finding more capable divers. However, this was impossible, so on 25 August the Kampen engineer carried out his own investigation. To his astonishment he found traces of gold on the steel tip of the sounding-rod with which he was endeavouring to investigate the lump of rust. Diving was therefore resumed on 31 August and hopes ran high. However, the diver came to the surface declaring that there was nothing to be found and saying so in tones which suggested he had given up.

The tongs were now the last resort. G. Bakker cast out the tongs and soon afterwards fished up a lump of rust ... not only a lump of rust but also the first louis d'or and the first Spanish piastre. On that same day a gun barrel was recovered and on 1 September two silver spoons.

The recovery of the first coins brought about a change of attitude in the divers. They suddenly made totally unacceptable demands and threatened to work at salvaging the treasure on their own account if these demands were not met. In fact, they twice sailed out to the site of the wreck, but only to find it occupied by Taurel and the members of the Terschelling board. Thereupon they came to an agreement with Taurel so that they were able to resume treasure-fishing with renewed hope and combined forces.

Although the men from Egmond gave no more trouble, the news that the loot evidently lay there for the taking brought competitors to the scene. As early as 18 September 'the vessel belonging to one of these conspiracies arrived on the site of the wreck, positioned itself without delay above the favoured spot and stuck to this position so stubbornly that it was impossible for our people to continue with the salving', reports the *Further Communication*. The log book reports that no salvage work could be done between 15 and 26 September because 'unathorized persons forcibly prevented the work from proceeding'. In a letter to Otto Eschauzier, written on Terschelling on 31 October, some sharp things are said about 'the scandalous intrigues going on here, the obstruction of all kinds and finally the rows this leads to'. The letter writer, Pierre, says: 'I had one yesterday, for instance, with Burgr S' (this was apparently A. Swaan, a burgomaster of Terschelling from 1837 to 1859, with whom, in his function as Chief Receiver of Wrecks, a financial agreement was soon signed) 'at the same time as Wilhelm was having one with those scoundrels from Urk, giving them a proper wigging.' He ends by letting out the very latest item of news: the contractors had put in an official complaint to the Dutch Home Office and Ministry of Justice about the behaviour of the burgomaster of Vlieland, who 'began promising 33 per cent of anything one should bring him from the *Lutine* and who is at present organizing a diving expedition with the notary Kikkert of Texel, a Jew, Van Gelder, and a diver, Bijl, from Den Helder, who is always drunk'. In another (undated) letter, Kikkert, a member of the provincial parliament of North Holland, is named and a certain Nijkerk, a solicitor from Amsterdam and second-in-command of the competing enterprise. When they attempted 'a poaching trip with their own ship and their own divers' to the *Lutine*, they were 'sent off by the military'. Indeed, after 1 October a gunboat had appeared off Terschelling at Eschauzier's and his colleagues' request to the authorities 'to safeguard the rights of the enterprise'. It worked, and we find Eschauzier writing with satisfaction: 'Frightened off by this intervention, the other fortune-hunters did not dare to carry out their plans and from now on the enterprise enjoyed exclusive and undisturbed rights of salvage in practice as well as on paper.' Moreover, this warship seems to have kept an eye on things in 1858 too, for according to a letter of 21 May of that year, the commander, J. C. Oudraat, of gunboat No. 32 *Terschelling* requested the burgomaster of that island to allocate him a warehouse for his crew's victuals, while as late as 6 January 1859 the Royal Commissioner had the burgomaster approached to ask why the vessel was still required 'in this season of the year'. The burgomaster replied that when sailing in and out of the inlet, many fishermen 'often damage and sail over the marker buoys for love of mischief or out of jealousy and may also cause the wreck to silt over with their drag-nets or trawl-heads', and that the presence of an armed vessel would keep them from committing such outrages.

In 1857 the work went on until 26 October, when fog put an end to operations for that year. By then about 20,000 guilders in gold coin had been recovered, most of it in piastres, of which Lloyds received half, Taurel 4,000 guilders' worth, the shareholders a like amount, and the Eschauzier heirs 2,000. While this result was not unsatisfactory, 1857 was to give the heirs another cause for satisfaction. The difficulties with competitors had again raised the question of who was authorized to carry on the salving operations. On 19 October the Home Minister approached the government attorney on the matter and on 2 December he had the royal commissioner for North Holland informed that he was in agreement with the recommendations he had received. This implied that as long as the authorization granted to P. Eschauzier on behalf of the association by the royal decree of 1821 had not been withdrawn, the association's right to search for the cargo could not be disputed. Any person attempting to recover goods from the *Lutine* without a licence to do so should be forced to retire, the minister wrote, a point of view which led to the signing of an agreement between Swaan and the *Lutine* enterprise soon afterwards.

The month of May 1858 was characterized by northerly and westerly winds which resulted in constant rough water. It was not until the last day of the month that the Egmond divers' boat was able to go to the wreck. The divers went down eight times, but to no avail. On 1 June they had better luck. The second time he went down Wijker found the first silver bar and four more were brought up that same day. Two days later the same diver recovered the first gold bar, and in the ensuing fortnight so much precious metal was raised that by 14 June 66 silver and 27 gold bars had been salvaged, 4 more gold bars being recovered in July. The helmet diver P. Swart was exceedingly lucky, bringing no less than 34 silver and 17 gold bars to the surface. They thought they had now struck the treasure chamber, but it soon became clear that what they had found was indeed a rich spot, but not the main store. At last the results became so poor that Taurel himself went down in a diving helmet on 18 June, repeating the venture on 12 August. But only piastres were found and the diver Swart accordingly expressed the opinion that they might as well suspend operations.

Nevertheless, they wanted to have one more try and a second vessel was allowed to take part, so that on 23 September there was a dual descent. But the new helmet diver was not happy, considering the obstacles so great that one descent was enough for him. That day one of the Egmond divers brought up a gold bar but it was to be the last of that year, 1858.

The following year hardly began hopefully. On the first day 7 May, six descents yielded one silver bar and two piastres, but during the whole of the rest of the season only four gold bars and one of silver were garnered. Of the 63 descents during May and June, 40 were abortive. Taurel and his colleagues could not and would not believe that the *Lutine* had yielded up all her treasure. On the contrary, they had no doubt that the greater part of the missing specie must still be lying somewhere close to the wreck. The extensive enquiry made – in which J. Reedeker, particularly, played an important part – led to a sketch of the wreck that is of great significance to our knowledge of the disaster. This sketch was printed in the *Further Communication* of 1861 and twenty-five years later was also included, with a few small modifications, in *Het Bergen van de Schatten uit de Lutine* (The Salvaging of the *Lutine*'s Treasure) by W. H. ter Meulen. We shall discuss this latter version on the basis of the 1861 date. The wreck lay from SSW to NNW, her surviving timbers lying below or level with the seabed, with some parts protruding above it. Since the ship had heeled to starboard, the broken-off timbers on this side were just visible here and there above the sand, while most of those on the port side rose six to ten feet higher. Two heavy anchors were found at the front of the wreck, where formerly the stem had been. The middle section was covered by a large mass of iron ballast, while further aft, the gold-containing rust having been cleared away, the keelson lay exposed. It was in the vicinity of this member, on the end of which the divers would sometimes sit to rest, that the helm was found, from which it was possible to deduce that the peak (the ship's treasure store) had been thereabouts. And it was, indeed, roughly at that point that the first bars were found. That the powder room lay there is certain, for a lot of ammunition and broken powder barrels with copper hoops were found.

The treasure fishers were not very interested in the cannon that were found, being far more intrigued by a separate, fairly small piece of wreckage about and beneath which the other bars were found. The divers had kept this find to themselves, but it had now been discovered and Taurel and his companions supposed that this piece of wreckage was the stern part of the ship in which the long-sought treasure chamber was thought to lie. Investigations in July and August 1859 confirmed this supposition, even though it was not terribly convincing as far as the size of the finds was concerned. After a few of the ribs with planks and copper sheathing had been pulled away, meticulous inspection did, however, show that they had belonged to the starboard side of the stern, and since this side had been neither removed nor fractured, the conclusion that the stern was still intact and that the port side of it had sunk into the seabed was inescapable. Thus the starboard side above it formed a sort of lid over the treasure chamber. When this cap was pulled away, another four gold bars and one silver bar came to light, while the number of piastres recovered exceeded 3,500. Yet no sooner had the lid been removed as far as the divers could manage it and the valuables salved, than it seemed the source of money had been exhausted. Report no. 106, made by Captain T. J. Reus on 15 October 1859 and to be found in a collection of daily reports in the municipal archives at Terschelling, reflects the disappointment encountered, whilst being a remarkable piece of maritime prose. It reads:

'Saturday 15 October, at 8 o'clock in the morning, bom (the fishing vessel of the men from Egmond) and a schuit (the Terschelling board's boat, exercising supervision) sailed to the *Lutine* with a south-easterly, arrived at the markers at 11 o'clock, moored the bom, fixed the ladder and Gerrit Wijker went down, was down for 20 minutes and came up owing to the incoming tide, reported he had been able to see or find nothing and said ladder had stood on the same spot as the previous times, had to wait then for the ebb tide, since we couldn't sail off. With the turn of the tide we raised anchor and sailed for the harbour, where we arrived at 8 o'clock and brought the report to the dupo' (read 'depot').

The men from Egmond saw no point in staying any longer and consequently left, promising to return in the spring of 1860. But in that year the results were so bad (something more than 200 piastres and no bars at all) that they ceased operations after 25 June and left Terschelling. That 25 June was not even one of the worst days appears from report no. 11, which also gives a good idea of the diving technique. In it Captain Reus testifies as follows:

'June 22, 23, 24 no opportunity to work. Monday 25 ditto. $3\frac{1}{2}$ hours in the morning board's vessel and fishing vessel sailed out of harbour, at $6\frac{1}{2}$ hours anchored at the markers of the *Lutine* wreck, the bom being anchored straightaway, it being low water. Ladder placed, the diver P. Swart went down. After 20 minutes down, he came up, having 88 whole 1 half 2 quarter Spanish piastres in the sack, of which 3 completely in a piece of rust, reported that he had not been able to see so well on the bottom, and that there had been only one small spot where he could work the reason being there were still two 'hoogtens' of sand on the bottom. 41 feet of water found with sounding-lead, took the finds on board the schuit (the board's boat) and waited for high tide. $1\frac{1}{2}$ hours high tide in the afternoon, ladder placed, diver G. Wijker descended. Being 45 minutes below, he came up, having 37 whole 2 half 1 quarter Spanish piastres in the sack and tied to the ladder rope 1 copper hoop, 1 three-sided ballast weight. The diver reported that he could see quite well on the bottom, 45 feet of water found with sounding-lead, took the finds on board the board's boat.'

Nothing more was found in the remaining months either, while at the end of October, a thick fog prevailing, they were not even able to locate the wreck, since the buoys had broken loose: a hardly triumphant conclusion to a series of salvage attempts which all in all had nevertheless yielded half a million guilders; precisely 41 gold bars, 64 silver bars and 15,350 gold and silver coins to a total value of 529,487 guilders, which figure includes a small amount recovered in one or two years after this. The enterprise was able to pay out 136 per cent on its shares!

Work did not come to an end immediately and Brand Eschauzier was still hopeful in 1862. His hopes were founded on an invention by Taurel which contributed little to the successes of the period around 1860, yet of which something should nevertheless be said. It was a diving apparatus mounted on a sailing ship. The design was Taurel's own and the first we hear of it is when Jean Brand Eschauzier and Louis Taurel came to an agreement with the shareholders. It concerned the *Hollandsche duiker* (Dutch diver). It was called Dutch 'in contrast to already existent English and American diving bells', wrote Brand Eschauzier, noting that the bell which the government 'had most obligingly put at the enterprise's disposal' was all but unmanageable due to the weight, position and arrangement of the crane, and had even proved highly dangerous, while there was far too little room for it.[31]

The invention took the form of a vessel specially designed for the purpose, with a diving bell and an air pump, to be driven by a small steam engine of 5 h.p. According to the agreement, the vessel *Hollandsche Duiker* was to begin work in June 1859 and invitations were, in fact, sent out during that month to view a number of tests to be held on the IJ at Amsterdam, near the head offices of the Rhine and IJssel Steamship Company.

The bill for the installation has been preserved. The bell with lead ballast, the body alone of which weighed about 6,500 lbs., was manufactured by the firm of Van Galen en Roest of Kampen and cost over 3,843 guilders. The air pump was made by Schutte Brothers of Kadijk, Amsterdam, and was entered as costing 3,500 guilders. The vessel itself cost 19,000 guilders and was built in the King William Yard of A. van der Hoog, Amsterdam. Thus a great deal of money was wrapped up in this experiment and friends of the Eschauzier family did not fail to warn them of the risk. As early as 2 March 1859 J. Reedeker wrote that if another bell had to be used, he would place it 'alongside rather than inside the ship' in order to be able in this way to offer assistance, if required, from outside. After which he went on: 'Be this as it may, we wish Mr Taurel success but urge caution upon the Eschauzier family in a matter which will entail terrific expense and in which they should not become involved as a participant (they already having a good share in the enterprise).'

It was Reedeker who on 18 August 1859 drew up a long critical report on the operations of the *Hollandsche Duiker*. He had established that the large bell was a danger to the helmeted divers, while the bell divers proved afraid 'to find themselves on or against the wreckage'. The funnel in the middle of the bell proved to be a great handicap as well, dangerous at times, and no digging could go on under the bell. He had no complaints to make about the crane, the air pump and the ship. But this always highly enthusiastic member of the board did have objections to the way Taurel and his divers went to work. On 13 August – when one of the men from Egmond had been in danger – they had again missed 'the right square', 'a small corner of the wardroom which has collapsed on to the bottom'. Therefore, he argued, they ought to open up the bottom itself instead of the sides, for 'in 1858, contrary to my advice, the helm was raised and there, in the resultant hollow, our fine hopes were lost forever. Let us learn from experience.'

These remarks were apparently heeded. In 1860, at last, Taurel appeared on the scene with a modified bell. The new *Hollandsche duikerschip* arrived in Terschelling harbour on 20 July, tests being made three days later. After the ship had shown that she 'turned and cruised fairly well', the diving bell was let down and pulled up again in the forenoon in about eight fathoms of water, 'and then found that all instruments satisfied requirements and in the afternoon the ship sailed into the harbour'.

Had Lloyds – whose appetite had obviously increased while eating – little faith in Taurel's new invention? In any case, on 27 July the board on Terschelling received a helmet diver's apparatus from London via Harlingen 'the which was immediately accepted by Mr Taurel on the Dutch diving ship'. On 30 July Heinecke, an Englishman, arrived on the island with two British diving colleagues. They took up quarters on board the diving ship that lay in the Schuitegat, where 'they tested the diving apparatuses which Lloyds had sent thither for use'. The British divers made their initial descent on 1 August. The first man reported 'that he could clearly discern all the objects lying on the wreck, that he had seen a mass of copper lying there, that he had been able to see the stern and timbers and supports, etc.' That afternoon the second diver went down on a short ladder that did not reach to the bottom. But this man soon came up empty-handed – he had not been able to see a thing. After this trial we find no further mention in the documents of activity by British divers.[32]

As far as the apparatus was concerned, in October 1890 W. H. ter Meulen, who made Taurel's acquaintance after the period of salvage just discussed and who had evolved a method of his own, testified that the diving bell was excellently designed and that the fact that it had been mounted on a sailing vessel instead of a steamship was not the inventor's fault but the result of a shortage of capital. Ter Meulen declared: 'Engineer Taurel deserves much praise for carrying out this work. But the bell could not be placed among the wreckage and it was difficult to dig the sand underneath it; it was accordingly of no use. Only the divers in helmets raised salvage to the value of 529,487 guilders.'

To begin with, however, Brand Eschauzier and Taurel could not be convinced of any shortcomings in the bell and were able to quote the opinion of the British diver Heinecke in support of their view. Heinecke declared that of all the diving bells he had seen 'there is none that is so efficiently designed and so perfectly complete as the *Hollandsche duiker*'.

For the benefit of the shareholders in the enterprise the two men calculated that their construction would have a working capacity forty-eight times as great as that of the fishing vessel (the bom) with its helmet divers, one more reason why they announced the continuation of activities in 1861. And so we see the contracts extended year after year. The last occasion was in July 1862, when Taurel was still on Terschelling. Although little is known about them, the attempts made with the *Hollandsche duiker*, captained by P. T. Krul, appear to have gone on during the entire summer of 1862. There is even a letter of 11 August to Mrs Eschauzier-Uitenhage de Mist, Pierre Eschauzier's widow, of Kampen, in which her son, underlining the word 'encouragement' twice, reports on an investigation the diver Drijver had carried out on the wreck with a steel pricker. He had discovered a hard floor about twelve feet square in which there were regular seams between hard objects one and a half decimetres across. At first he thought they were ballast but was later convinced they must be silver bars, for the ballast pigs were invariably stuck together with rust. Moreover, the surface was soft and when Drijver came up again his hands were tinged blue. So, full of hope, he went down again the next day, armed this time with a pair of tongs and a pick. He found the same floor, but this time it was buried under such a mass of sand ('a herald of bad weather') that he had to give up. That same evening the storm blew up.

All the same, the divers were still in very good heart and Krul had gone down and convinced himself of the accuracy of Drijver's report. But in a letter of 7 November from the board on Terschelling to members of the board of the *Lutine* enterprise in Amsterdam, nothing remained of this optimism:

'This week Mr Taurel has been over the *Lutine* for days and nights in the board's vessel and both Drijver and Krul have been down repeatedly, but with a great deal of trouble from dark water and turbid sea, nothing of value found ... We do not yet know whether Mr Taurel has decided to give up altogether for this year but the expense is heavy and would have caused many in his position to have abandoned the contract long ago.'

The anticipated end to the proceedings was not long in coming. Although on 25 April 1863 Brand Eschauzier sent round another circular in which 'perseverance with the work' was said to be 'imperative' and which also included a reminder that 'numerous diving personnel with the requisite vessels and tools' had to be continually on hand in the harbour of Terschelling 'in order to work on the wreck immediately the first opportunity arose', a month later this episode had also become a thing of the past. On 21 and 22 May the parties concerned signed an agreement whereby the contract of 5 August 1857 came to lapse as from 1 June, since Taurel wished to settle abroad. Brand Eschauzier assumed the obligation single-handed to 'proceed with the salving without delay with tools and vessels he himself should supply and finance', in return for which he was to receive 15 per cent of what might be recovered.

The result? The last document in the *Lutine* archives on this particular matter is a letter from Lloyds' Amsterdam agent John Mavor Still, who wrote to Captain Krul on Terschelling on 21 July 1863 to say that in his opinion nothing of value was left on the surface of the sea bed – a fairly obvious conclusion after three years' fruitless effort – and that they should consequently take an example from 'various technical persons in England and elsewhere', who had had great success with placing barrels of stones and fascines on the sea bed 'which causes a severe friction'. And he added that Lloyds – which shortly before had received a quarter of a million guilders without having had to spend a penny – wanted these methods employed as well. They should not, according to this agent – who also supplied a detailed calculation of the amount of treasure on board the *Lutine* –

'spare themselves any trifling expense and trouble, especially as this could lead to good and desired results'.

Did this encouragement – which did not cost Still a penny either – bear fruit? In any event, Captain Krul enjoyed no advantage from this further work on the *Lutine*. As late as 1868 we find him signing a receipt by which, as former captain of the ship known as the *Hollandsche Duiker*, he accepted from the authorized representative of the owners, J. P. Brand Eschauzier, fifty shares for all his claims, amounting to 3,000 guilders, the said shares having been ceded to him by the heirs to Pierre Eschauzier's estate. What is more, he was to pay all eventual outstanding claims by the inhabitants of Terschelling, whether workmen or suppliers, out of his own pocket. Krul was not to be the only one who had to pay in disappointment and hard cash for the faith he had placed in the success of the treasure-fishing. But yet another candidate for the rôle of victim had come forward.

16 W. H. ter Meulen in the Tease's thrall

'In this essay – the outline of a plan for setting to work on the *Lutine*, regarded exclusively from a technical viewpoint – I have said nothing of the millions ... Readers who prefer to hear about that are not the people for me. I am looking for kindred spirits, who consider sanddiving an absorbing business.'

The man who wrote these lines, Willem Hendrik ter Meulen, may be termed a victim of the fatal ship, a man who was enthralled his whole life, not so much by the British frigate's cargo as by the possibility an invention of his own offered for recovering it. A man who, at the end of his life, after many sacrifices of time, money and energy and after many disappointments, still had not had an opportunity to demonstrate this possibility, and whose life, as far as this principal aim was concerned, must accordingly be regarded as a failure. A man, finally, who looked to the historian for vindication: 'If ever the history of the *Lutine* is written, people will think better of me than they do at present.'

Who was this Willem Hendrik ter Meulen? If he himself had not enabled us to get to know him by a veritable avalanche of letters, brochures, remonstrances, reports, addresses and apologia, a member of his family, F. P. ter Meulen, wrote in 1907 a booklet which was not put on sale to the public.[33] We have borrowed from this work, especially as regards a few details known to insiders only.

Willem Hendrik was born at Bodegraven on 11 January 1830, the son of a small manufacturer of white lead and vinegar. The boy showed an enquiring mind at an early age, but suffered from deafness from the age of ten as a result of measles, and a few years later he was stone deaf. There could therefore be no question of further study after elementary school and he had to look for something to occupy him at home. His father's factory offered an outlet of a kind, but work there could not satisfy his lively and industrious mind. He soon developed a passion for natural science and became keenly interested in things electric and mechanical. His nephew, H. Enno van Gelder, himself an engineer, consequently praised him after his death for being a 'born engineer', who had 'invented and registered highly ingenious affairs' in the way of diving apparata, shell-sorting machines, buoys and the like.[34] At the end of the account of his life, his biographer wrote of him as being of 'thoroughly good heart, uncommonly lively mind, and firm and strong character'.

How this gifted yet handicapped young man came into touch with the work on the *Lutine* is not clear. During the summer of 1859 he was staying on Terschelling. He did there what many another visitor does: 'standing on the beach, he moved his foot about in the sand and noticed how soft and porridgy it was. It seemed as if the water that welled up in it in response to pressure at the side made the sand fluid. Couldn't one, he wondered, easily penetrate the sea bed by forcing water into it?' Ter Meulen, then twenty-nine, constructed and patented an apparatus to make the sand 'fluid' in order to be able to clear it away. It consisted in the main of an iron tube out of which water was expelled through numerous small holes, whereupon the sand would be carried away automatically by the current, and was called a 'zandblazer' (sand-blower). In 1859, on Terschelling, he made the acquaintance of Taurel, who was also staying on the island at the time in connection with the salvage operations. The Kampen engineer was interested in the co-operation Ter Meulen offered, whereupon the inventor had one of his sand-blowers constructed at his own expense. Its manufacture did not turn out to be all that simple but at last it was ready and plans were made to test it out on board the *Hollandsche Duiker*. When, however, the inventor arrived on the vessel, the necessary arrangements had not been made. Neither did the board of the *Lutine* enterprise wish to commit itself to the payment of any remuneration or grant any financial assistance.

We have already seen that Taurel abandoned all his attempts in 1863. He left for the Indies and six years were to pass before he and Ter Meulen were to meet again – above the wreck of the *Lutine*. Full of energy, the unmarried, deaf inventor from Bodegraven was going ahead with his research. Lacking all training, all contact with men of learning or with anyone at all but his nearest relatives, he worked out his highly promising plans in the quiet house in the still quieter little town in the province of South Holland. He became the inventor of what was at first called a 'zandboor' (sand-drill), and later described as a 'sand-diving apparatus'. It amounted to a device which forced water into the sandy sea bed in order to clear a way for a helmet diver. The merits of this Ter Meulen went on propounding to the end of his days, with enthusiasm, with stubborn conviction, but without success.

To begin with, the inventor was not successful among the scientists either. His first encounter with Stieltjes, the well-known engineer, and Van Diesen, the engineer who had built the railway bridge near Culemborg, was disappointing. Both men came forward with counter-arguments and adopted an attitude of reserve, though later on Van Diesen was to defend the concept of sand-diving – a far from happy term from the psychological point of view. All the same, as early as 1863 the inventor suggested the idea of sand-diving to the *Lutine* enterprise, which had been in existence since 1821 and which in his writings he always called 'the Dutch fishery at the *Lutine*'. It was not until 1867, however, that they put him in a position to purchase and equip a steamship from which the diving was to be done. In April of that year he signed a contract, of which he said later: 'From then on I devoted myself to the *Lutine* affair as if I were the concessionary and not merely a contracting party for a limited number of years. My task was a dual one. I had to choose a system for setting to work on the *Lutine* that would be satisfactory in every respect and I had also to find out where the wreck lay.' By the terms of his contract he was obliged to maintain a careful watch on the *Lutine* in order to find out how much sand lay on top of the wreck. As soon as the water above it was about twenty-three feet (seven metres) deep, the now patented sand-diver would have to start work and descend another six or seven metres through the sand to reach the wreck. But the contractor could start to work before this, if he wanted to. In any case, he was to take official soundings of the wreck at least three times a year, at which soundings representatives of the *Lutine* enterprise and of Lloyds were to be present, while he was on no occasion to proceed to the site of the wreck without first having informed those concerned. The contract was signed for a three-year period and now Ter Meulen could act in public as a contractor. He sent out a circular raising a loan of 300,000 guilders so that 60,000 would be available to the enterprise. This money was raised to defray the costs and to attract capital to redeem the loan. The redemption and interest (5 per cent) payments would be made from an inalienable deposit of American funds. In a booklet Ter Meulen informed the public in 1896 that this type of loan had been chosen in view of the *Lutine*'s bad reputation. The shareholders wanted more certainty of getting their capital back soon and also wanted to be assured of receiving a fair rate of interest, 'which requirements have been fulfilled to the letter,' Ter Meulen was able to write with satisfaction afterwards.[35]

In 1867 he left for London, where he purchased the *Antagonist*, an English steam tug driven by paddlewheels and constructed of steel. Ter Meulen so modified its 50 h.p. engine that it could be disconnected from the paddle wheels and connected up to the centrifugal pump which 'by employing very heavy reduction gear makes thirty revolutions to every revolution of the engine's crankshaft'. The centrifugal pump, known as a 'whirlpool' pump, lay horizontally below the waterline in the steamship's fo'c'sle, so that it was always full of water.

Among her equipment the *Antagonist* had a hydraulic press for pulling away pieces of wreckage, a derrick aft to haul loads up out of the water and drop them into lighters, and an apparatus for the underwater ignition of 'lithofracture'. 'The shattering power of this material, which is fortunately harmless in use, is wellnigh unimaginable,' Ter Meulen wrote proudly. This 'lithofracture' had been used shortly before with great success for work on the steamship *Telegraaf*, sunk off Willemstad in the Antilles, making it possible to salvage the boilers (which had already been written off) and a fine crankshaft.

One of the men who entered Ter Meulen's service was P. T. Krul, previously the master of the *Hollandsche Duiker*. But not much could be accomplished in 1867, for in December 1863 the Zuidkaap marker on the Noordsvaarder sandbank had been blown away in a storm, while the wreck buoys had disappeared long since as well. Fortunately Taurel, who had been working in Java for some years, put his old notes and sketches at Ter Meulen's disposal during a meeting with him in the autumn of 1868 and after this the two worked in close co-operation. Although the soundings could now take place, the results showed that for the time being no salvage attempts could be contemplated. In 1869 there was, on average, eight feet of water above the wreck; in the months of May, July and September 1870 twelve feet; and on 7 July 1871 as much as fifteen feet. But in 1872 this figure had shrunk to seven feet again and in 1873 to six feet, minimum depth, which meant the sand on top of the wreck was at a maximum. All the same, now and again the backers of the enterprise had a chance to see what Ter Meulen's invention was capable of.

This invention took the form of a drill consisting of a kind of grid with a circular base. It was manufactured of sheet copper (which is less liable to corrode in salt water than iron) and into it could be forced water, which had no outlet other than the hundreds of small holes in

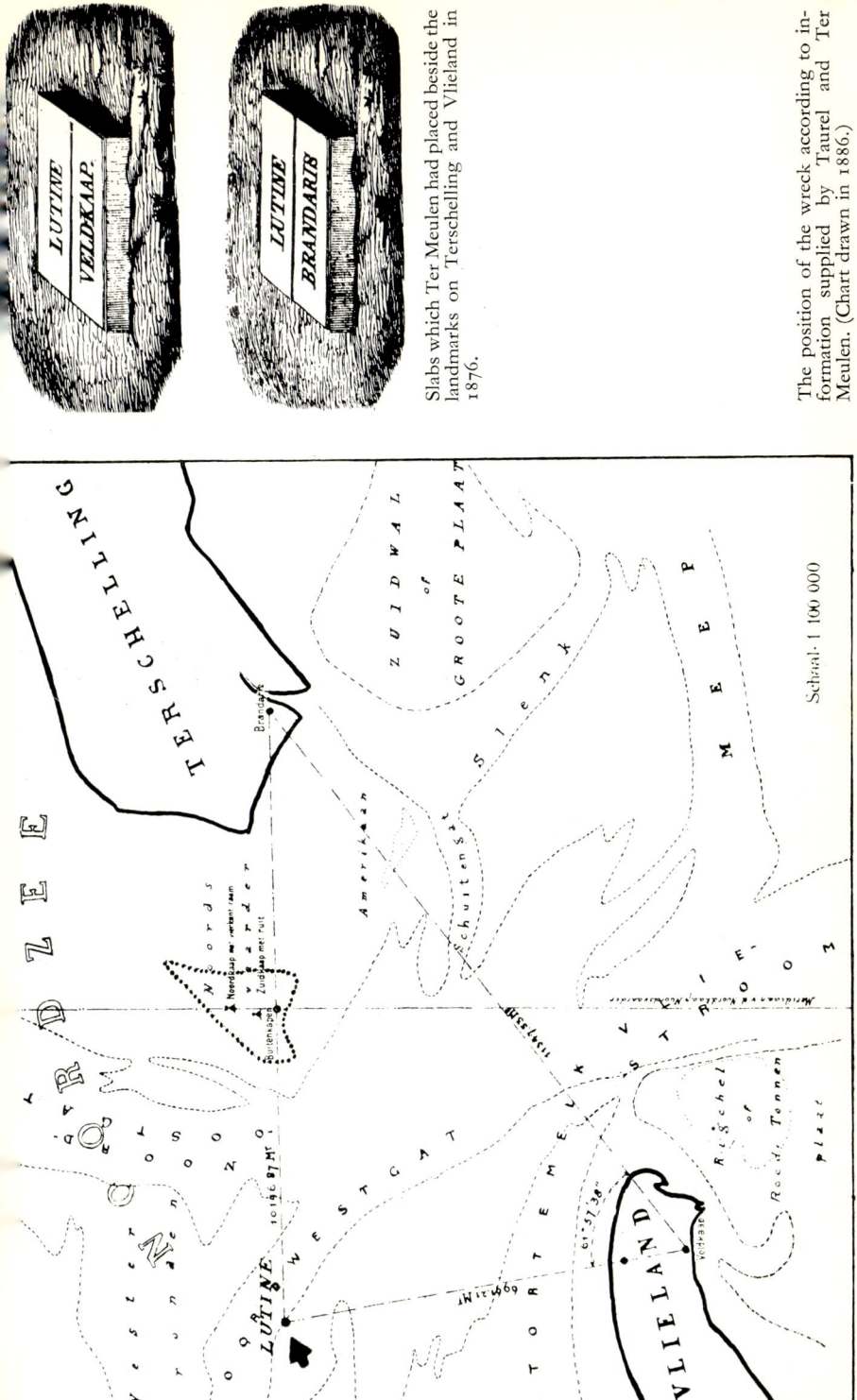

Slabs which Ter Meulen had placed beside the landmarks on Terschelling and Vlieland in 1876.

The position of the wreck according to information supplied by Taurel and Ter Meulen. (Chart drawn in 1886.)

The British lighter *Bill O'Malley* with a bower anchor raised in August 1911. On the right, the *Lyons*.

the bottom of the hollow grid placed flat on the sea bed. The water was pumped via a tube attached to the centre of the grid, being forced down by the steam-driven centrifugal pump on board the *Antagonist* at the rate of twenty-eight cubic yards a minute. But tests soon showed that approximately two cubic yards a minute was more than sufficient pressure to remove the sand. Next to the drill there was a probe which could be used independently to search for a concealed wreck. This took the form of an iron pipe weighted with a covering of lead, about four yards in length and attached to a separate water pipe of much smaller diameter than the drill. Not only did a jet of water spout out of the probe, but the jet opened up a path in the sand for the heavy pipe, which was only two inches in diameter, and the wet tubing slid so easily through the narrow hole in the sandy floor after the iron pipe that within a couple of minutes the probe had penetrated twenty-five feet, striking against the wreck's timbers. The contact could be detected on board the *Antagonist* from the vibration it set up. The tell-tale was an ingenious device on this probe. It amounted to an electric mouthpiece which reported the presence of silver or gold, i.e. non-ferrous metals. Two small electrodes with white metal tips were linked by telegraph wires to a small battery on board the *Antagonist*. Should both of these feelers touch the same piece of gold or silver this would complete the circuit and set off an electric buzzer.

Those who approached Ter Meulen's invention with some reserve were no doubt convinced of the tell-tale's efficacy. What they were worried about was the danger the diver would be in when he descended into the cavity that had been bored. Wouldn't its steep walls collapse and bury this seeker for gold under the sand? The inventor never tired of showing by demonstrations and by statements by his divers that this fear was unfounded. The log book kept by the board of the enterprise contain an interesting report of one such demonstration. We read:

'16 Sept. 1870. Riding at anchor in three fathoms of water in the Schuitengat [a navigable channel south of the Noordsvaarder sandbank just west of Terschelling] the boring machine was let down into the sand to a depth of fifteen feet, as was ascertained by the sounding-lead. After this the diver went down to the same depth, remaining in the cavity bored for about half an hour, coming back to the surface again with the machine without encountering any difficulty: this diver was a very calm and modest man and these tests convinced us that the salvage system on the *Antagonist* works efficiently, the which leads us to anticipate a favourable result once the level of water above the *Lutine* is more suitable.'

This was the view of Messrs P. Altena, D. Reedeker and W. L. S. Eschauzier, who a month later were writing full of enthusiasm to their colleagues and fellow board members in Amsterdam of their desire 'to keep Mr Ter Meulen associated with us with his diver, whose like we have not found before'. Yet seven years later demonstrations were still required to prove the invention's usefulness. The *Algemeen Handelsblad* of 10 December 1877 carried a report of tests near IJmuiden 'with the co-operation of L. Taurel, the well-known engineer'. These tests had been done within the harbour basin on 19 November, in the North Sea Canal on 20 November, and again on 5 December in a place not specified. In the last instance it was not a case of sand-diving but apparently of marsh-diving. An experienced diver, one J. Vrouwes, lent his co-operation for the purpose. He was hung about with so much lead that he weighed over 35 stone. The report goes on:

> 'Thanks to the design of his diving suit, the lead did not weigh down on him and it was possible to wind him down as he clung to a chain. He undertook his trip through the dark in good spirits and without anxiety, and when he came up again he was sprinkled with sand, which had stuck mainly in the folds of his suit. He said that the sides of the pit were far more irregular than those he had formerly seen in pure sand. The strata of peat had acted as shelves and the sand underneath it had collapsed in some places ... According to him, diving in pure sand is far less dangerous.'

This stunt seems, too, to have been the highest achievement a capable and courageous diver was able to accomplish using Ter Meulen's invention. No wonder that, according to the newspaper, all eye-witnesses were 'very satisfied with the results obtained'. Among these eye-witnesses were the government engineer K. van Rijn, the harbour master J. F. Nieuwenhuisen, and N. Witsen, one of the directors of the enterprise.

Ter Meulen already had more successes to his credit, even if not with the *Lutine* – from which the sand obstinately refused to budge. But by the time his contract with the *Lutine* enterprise was renewed for another three years in 1870, he had already received open support from such well-known figures as Van Diesen, who had recommended his method at a gathering of the Royal Netherlands Institute of Engineers, Professor Matthes, who had done the same at the Academy of Science, and J. A. van Eijk, who gave a talk illustrated by demonstrations in June 1871 and who was soon to write an article in the magazine *De Volksvlijt* (National Industry) on diving in sand and the combating of sandbanks.

In 1871 the inventor from Bodegraven handed over the management of affairs at the wreck to Taurel, while Captain Krul continued to be master of the *Antagonist*. At the end of 1872 Ter Meulen obtained a twenty-year contract and came forward with a plan for a new loan: 'the watch on the *Lutine*'. After some difficulties of a juridical nature with the *Lutine* enterprise, a solution was found by granting Ter Meulen and his assigns the exclusive right to work on the wreck with the sand-diver. Before long the contracting party made the acquaintance of J. O. H. Arntzenius, aide-de-camp to Prince Hendrik, the brother of William III of the Netherlands. Ter Meulen succeeded in winning over the aide-de-camp to his cause, and in April 1874 an exhibition was held in the Palace of National Industry in Amsterdam at which a working model of the sand-diver was put on display. This convinced the prince of the great value of the invention, so that floating a loan of 200,000 guilders in the form of shares 'to be issued in four series, each of 50,000 guilders, at five year intervals and paid up at 10,000 guilders per annum', proved a simple matter. When Prince Hendrik asked for some shares himself, none were left, so some of those already issued had to be recalled.[36]

The inventor-cum-contractor, now enjoying support from all sides, kept the interested parties informed by means of annual reports. Thus the report for 1874 entitled *De Wacht en Visscherij op de Lutine* (The Watch and Fishing on the *Lutine*) contains interesting information about the pinpointing of the spot, which Taurel had to work out. The old marker on Terschelling no longer existed. The stormy tides of 1863 which swept over the Noordsvaarder sandbank washed away all the landmarks, the two colossal 'national markers' (Rijkskapen) and the north and the south markers, their replacements being sited at other spots. But the *Lutine* marker on Vlieland was still in place, while pinpointing the wreck's position was further facilitated by General Kraijenhoff's triangulation, and also by a map and description by A. R. Blommendaal and by some information supplied by J. Gorter, Ministry of Works superintendent on Terschelling.

Ter Meulen reported: 'The chief result of the calculations is that the *Lutine* is lying at a spot 53° 21′ 33″ 974 North and 0° 10′ 41″ 804 East of the Amsterdam meridian (the steeple of the Westerkerk). The great precision of these calculations will be appreciated when one recalls that where the *Lutine* is lying, one second of longitude represents 18.48 metres and one second of latitude 30.9 metres. In the calculations themselves they have even gone to thousandths of a second and the place pinpointed is that which during the operations in the years 1858–60 came to be taken as being the stern of the *Lutine* where the bullion lies.'

It was now a matter of erecting new, clearly visible landmarks covering both the Noordsvaarder sandbank and Vlieland, to which Minister for the Navy agreed on 4 September 1874. Little remained of the marker on Terschelling after a storm during the night of 21 October tore it from its moorings. Later it was found along the coast of Friesland, near Sint Annaparochie, still whole and resembling a small building about twenty feet tall. A new landmark was erected, and another on Vlieland. Anyone at the point where the bearings of the two markers intersected at an angle of 80° 46′ 12″ 36 knew for certain that he was above the celebrated wreck.

In 1876 stone slabs were laid beside both landmarks and anchored to an iron base going down to a depth of three feet. There was no possibility, therefore, of their changing position. The slab on the Noordsvaarder was marked LUTINE BRANDARIS and that on Vlieland LUTINE VELDKAMP. In the meantime, soundings were continuing. In 1876 the third, or autumn, sounding was taken in ideal conditions and showed a fairly large improvement at this time, though the depth of five metres, which by the terms of his contract would oblige Ter Meulen to place a buoy, was not yet in sight, still less the seven metre depth, when the diver would have to start operations. In fact the depth of water above the wreck was so small that the *Antagonist* was unable to anchor there, and a sailing boat was used for taking soundings.

When, in 1879, the members of the board were invited to provide another loan, only one fifth of the amount required was subscribed. The names of certain prominent personalities who gave the loan their blessing proved insufficient inducement. Ter Meulen also had bad luck, for both Prince Hendrik and his own brother, Arent, who had written a publicity brochure for him, died. Neither the two exhibitions in the capital at which a model of the sand-diver was put on display, nor the speeches delivered by two engineers, C. van Hasselt and K. van

Rijn, at the Royal Institute of Engineers on 3 July 1878 had the desired effect. When, in addition, Ter Meulen suffered other financial setbacks unassociated with the *Lutine* affair, his days as a salvor were numbered. He was obliged to dismiss both Taurel and Vrouwes, the loyal and skilful diver, and although he kept the *Antagonist* as long as he could and even went on taking soundings with her, it was not long before the ship, which was costly to maintain, had to be got rid of. Ter Meulen continued, however, to be the contracting party, still hoping that sooner or later the financial means would nevertheless be found to investigate the wreck with his sand-diver when the water was deep enough. So he went on using his own funds, which, his biographer writes, were very limited, taking soundings with a boat he had to hire. The bullion itself, however, was of little interest to him. He himself wrote: 'Were I to haul up gold on one side of the boat, I should be capable of dropping it into the sea again on the other' – hardly a tactful remark for the ears of those who were out for monetary profit first. But what concerned him most of all was 'the use of his invention, the work itself'. His biographer also wrote: 'Any eventual profit had to be seen as a means and take second place. We may look on that as a virtue but to many such an attitude shows weakness.'

These words were written in 1907 but they had proved true long before. While the inventor from Bodegraven went on faithfully and hopefully taking soundings of the depth of water above the *Lutine* in accordance with the terms of his contract (on 14 November 1884 it had increased to seventeen feet, a depth not encountered since 1868), a shell-dredger loomed on the horizon. This was to mean the beginning of a new – though not the last – phase of the salvage operations, while to the man who for more than a quarter of a century had devoted his ingenuity, labour, time and money to the enterprise it meant the beginning of the end. The contention that it was now possible to raise the gold and silver with simple and economical means, caused the board members' thirst for gold to revive. Despite repeated warnings and against his will. Willem Hendrik ter Meulen became involved in a method which promptly led to the débâcle he had predicted. That he had to bear the main blame for its failure was only to be expected.

17 The 'lamentable era' of the shell-fishers

On 14 November 1884 W. H. ter Meulen found seventeen feet of water over the *Lutine*, and on 7 November in the following year, sixteen. But on 4 May 1886 the soundings showed seventeen feet again, and in the beginning of 1887 and 1888 as much as eighteen feet. The depth required to make use of the sand-diver's capacities was rapidly approaching. Was it to be used at long last? The *Lutine* demon had decided otherwise. On 28 April 1885 Ter Meulen received a letter from the head of the Dros Steam Shell Fishery of De Cocksdorp on the island of Texel in which this man declared he would be able to suck the bullion ship free of sand by using his suction shell-dredger *De Tijd*, and that he was ready to attempt the same, given favourable conditions, if they could show him precisely where the wreck lay. We need hardly say that Ter Meulen was not very enthusiastic about sand-dredging, much preferring to use his own method of sand-diving. However, there were members of the enterprise who expected wonders of the shell-dredgers: two, by the way, for in 1885 the steamboat *Friesland* put in an appearance. It belonged to Maas, a shipowner from Makkum who collaborated with Dros. The board members resident on Terschelling had had a talk with these shipowners and one of them, P. Altena, wrote enthusiastically of the unanticipated possibilities in a letter to the board members in Amsterdam N. C. and W. J. Witsen that:

> 'The desirability of Mr ter Meulen's entering into a contract with them is beyond dispute, since the method employed by suction shell-dredging is so unsurpassed in its practicability that it would border on frivolity were one to hesitate to grasp with both hands this single means of quickly recovering the *Lutine*'s treasure, amounting to about thirteen million.'

To frustrate Ter Meulen's objections Altena reminded them that in the late summer of 1880 the inventor of sand-diving (which, by the way, he now condemned as being 'wretched, dangerous and impracticable') had put to sea at Makkum to watch *De Tijd* at work 'and expressed his astonishment at the results'. That fishing for treasure was not the same thing as dredging for shells apparently escaped the writer of this epistle. In any event, he impressed upon the Witsens that Ter Meulen should sign a sub-contract with the Maas and Dros partnership.

Ter Meulen himself, in the meantime, had proposed that they should get his sand-diver working and fit it out with a Hall's injector. They could then begin in May, provided the requisite funds were forthcoming. He was thinking along the lines of £1,000 subsidy that Lloyds should make available, free of interest, receiving the money back out of the first remuneration due to Ter Meulen for his salvage operations. But Lloyds were not inclined to make such an advance. Their agent in Amsterdam offered Ter Meulen 10 per cent of the eventual yield... 'to enable him to co-operate with the shell fishers'. But this was not Ter Meulen's intention; he was not at all willing to commit himself to co-operating with dredgers with his sand-diver and therefore went on raising objections. All the same, in 1886 he gave in. His contract was renewable annually and he was afraid that if he continued to refuse his co-operation, the enterprise might annul their agreement that September. He also thought it probable that Maas and Dros would soon realize 'that dredging pits *on* the *Lutine* was not the same as dredging the *Lutine* free of all sand.'

Thus, on 12 May 1886, the Frisian Steam Shell Fishery of Makkum (Willem Maas) and Albert Dros and Company's Shell Fishery of Den Berg signed an agreement with W. H. ter Meulen 'in order that, working in association, they shall bring the *Lutine*, and her cargo included, to light, to which end the two steam Shell Fisheries shall make available the two steam suction dredgers *Friesland* and *De Tijd*, with their normal crews, and Mr ter Meulen shall provide the sand-piercer (*zandsteker*), two divers and two diving apparata.' The contract was for three years but would be automatically renewed for another three, should the requisite depth of water above the *Lutine* be maintained.

In July Ter Meulen transferred his responsibility to provide divers to L. Nieman,[37] a diver of Ouddorp, for 2 per cent of the yield. The man was soon to regret it. Just before this, on 28 June 1886, the first attempt was made with the small sand-drill, but without result. Bad weather brought postponements, but on 10 August the sound of the drill making contact with the wreck (and apparently with iron) was clearly heard on board the steamboat. So, as soon as conditions allowed, on 13 August, both dredgers were set to work to suck out a large hollow. The objects that came to light, such as lumps of rust with gunpowder and bullets, copper bolts and nails bearing the broad arrow mark, soon showed that they were over the right spot. On 20 August the first Spanish piastres were recovered, and a gold ring. This – after 87 years – added one more name to those of the few victims already known, for round the setting where once a stone had been, the words 'Sacred to filial love' were engraved, and on the inside 'Love and live happy I Hardcastle ob(it) 5 may 1795 old(d) 67 – E.H. ob(it) 26 july 1798 ol(d) 73'. The ring was thought to be a memento of the parents of one of the naval officers on board the *Lutine*.

There are several accounts of the way these and other articles were recovered, the most entertaining of them being by P. J. Wichers of Terschelling, who published a report in the Harlingen newspaper *De Echo* of 15 September 1886:

> 'According to the diver, the water is so clear here that, the weather being fine, he is still able to make out fairly well what is lying on the bottom of the hollow or pit, a thing definitely not possible further south. This [dredged] pit is shaped like a truncated cone turned upside down, at least forty meters broad at the bottom. The diver moves about on a floor covered with all manner of objects, cannon balls, pieces of wreckage, lumps of ballast, articles of various kinds and a great deal of rust, in which so far numbers of Spanish piastres and some pieces of gold have been found. The fact that the wreck has completely disintegrated makes searching exceedingly difficult. The masters of the vessels make little of this, however, since they are able to suck out a fairly large hole in a short space of time, which must be true, since they have already removed more than three hundred thousand cubic meters of sand next to the wreck.
>
> 'The diver can hear absolutely nothing in his suit. One can picture his position: he walks round, well harnessed, and searches about everywhere over the floor of a pit about fifty to sixty feet deep, surrounded by a wall of sand which can collapse at any moment, swathed in half-darkness in a silence as of the grave, and above his head the ever rolling, surging and swaying billows of the North Sea.'

Given the circumstances, Ter Meulen seemed reasonably happy about the way the shell-dredgers were working, even when the operations had entered their second year. On 22 September 1886 he wrote to Lloyds' agent that 'the team of workmen are excellent. They are tough fellows, men of resolution, especially Captain Van der Wal and Nieman and Sperling, the divers. I have great admiration for them.' Almost a year after this, on 11 August 1887, in a letter to the *Handelsblad* he declared he had been a witness to the great power of the two steam suction boats and to the contractors' perseverance. We do, however, discern a hint of criticism in the words: 'A wide pit has been dredged down into the wreck, which has been done after deliberation; for laying aside the possibility of an *early* yield of large proportions, they have taken the trouble to tackle the big wreck to begin with, in order to be surer of getting to the stern, where most of the money lies.' Yet he ends his account on a fairly optimistic note: 'If the weather is with us, there is every possibility of great valuables still being recovered this season—A large haul would be a just reward for the enterprise's untiring perseverance.'

Such remarks are in striking contrast with what Ter Meulen had to say in later years about this same activity. For instance: 'in the very first year the divers said they could not get into the irregular pits; they were completely discouraged even at that stage. Yet the lamentable toiling and moiling with the dredging went on for six years.' Or this: 'Must I undertake the new job with the sand-diver in such an unsatisfactory manner? What am I to do with these boats? It's as if a stable keeper should offer me his carriages just to get a percentage of the *Lutine*. Money is required to equip one of the boats for sand-diving.' This harsh criticism of the shell-dredgers and of the *Lutine* enterprise's insistence that sand-diving should be combined with dredging, to be found in a letter of 21 November 1889 to Van Buren Schele, a member of the board, is an allusion to a conflict that had arisen shortly before and had nothing to do with the poor results achieved, especially in 1887 and succeeding years.

A few objects of some significance were recovered during the first year. From 20 August to 9 October these were: 3,573 gold and silver pieces, including 3,304 whole Spanish silver piastres (worth 8,232 guilders and 10 cents in all), and a large number of small objects. The catalogue of the exhibition of finds organized in the *Buiksloter Tolhuis* (toll-house) at Amsterdam in November 1886 showed that the harvest during the late summer of 1886 was a fairly rich one. The catalogue arranged them under the following heads: armaments, navigation and ship's carpentry, domestic and personal items. Naturally there were cannon (including a 36-pounder still fully loaded and weighing approximately 1,600 lbs), cannon balls, grenades and pistols, and also boarding axes, two boatmen's breastplates, uniform buttons and the sheath of a dagger. Under the second heading there were *inter alia*, part of a sextant, the stern part of the ship, pigs of ballast, and a veritable arsenal of carpenter's tools such as chisels, hammers, an axe, a maul, a complete chest full of materials, lightweight chains and even three padlocks. The items in the last category were even more eloquent and included chest handles, a set of silver and pewter tablespoons, a fragment of a silver watch, a gold signet ring, an earthenware pipe, a seal bearing the name T. Ellis, a freemason's badge and a pair of stag's antlers, said to have been used as a coat-stand in the ward-room.

Although no bars had been recovered, the results achieved in 1886 made all – Ter Meulen included – hope for still better things. The situation at the wreck gave some reason for optimism according to a report from Terschelling, dated 14 June, in the *Handelsblad*: 'Today work was resumed on the *Lutine*. An area forty metres square has been marked out by four buoys and this area will be dredged and investigated to begin with. Soundings showed a marked improvement in the situation. Where there was formerly eight to nine feet of water, twenty feet has now been measured, so there is enough room to get rid of the sand they dredge up without any danger of their getting stuck fast in it themselves.' It was soon clear that this improvement in conditions did not mean greater success. According to Ter Meulen there was still too much sand lying on top of the wreck to permit them to speak of *regular* investigations. The haul bore this out: in 1887 the total amount recovered was valued at 2,174 guilders and in 1888 at 720 guilders.

These disappointments hardly fired them with enthusiasm. Whereas in 1887 Maas and Dros still wanted to carry on, while the masters of the two vessels, A. J. van der Wal and R. Veldhuizen, and the divers had already had enough of it, by 1888 all of them had become prey to despair. Nieman, who the year before had 'earned' a mere forty-three guilders plus a few cents, was asking to be released from his contract, and his men, Matthijs and Wouter, who had also been engaged on a part-share basis, were asking for fixed wages. (Nieman's total receipts for the 1886–88 period had amounted to little more than £22.) Even Lloyds' agent was forced to realize that dredging was a waste of time and money.

It seemed that Ter Meulen might get a fair chance to try out his sand-diving method during this phase. At the beginning of 1888 Figée, an industrialist from Haarlem, declared himself ready to grant him a steamboat so that he could go to work on the wreck alone. But this did not suit Maas and Dros, who saw no use in any method other than a combination of dredging and diving. The bickering went on for so long that Figée finally withdrew his offer. Ter Meulen made preparations for yet another demonstration of his apparatus and at the same time tried to terminate the contract with the shell-dredgers, having more success with the first aim than the second. The test was carried out on 21 October 1889, and the divers and the invited guests declared themselves to be very satisfied indeed.

Results there were none, for the time being at least. Maas and Dros had Ter Meulen informed by writ that they laid claim to an extension of the contract of 12 May 1886 and that they were counting on divers again in June 1890 – men whom Ter Meulen was liable to provide. Ter Meulen replied by writ that the shell-dredgers had failed to secure the required depth and that he therefore regarded the contract as having expired. Referees were appointed to settle the dispute, but they took a year to reach their verdict. By deed of compromise, dated 30 November 1889, Theodoor Stang, director of The Hague Water Board, and Amsterdam solicitors Jeronimo Catharinus de Vries and Menso Johannes Pijnappel were asked for a legal decision. In this decision they noted that a dispute had arisen in the spring of 1889 as to whether the agreement of 12 May 1886 had been terminated on 12 May 1889 (as Ter Meulen held), or whether the second three-year period had already commenced (as the plaintiffs held). Ter Meulen was found to be in the wrong and was obliged, with material and two divers, to resume work immediately with the plaintiffs' steam-dredgers and personnel.

Thus in 1891 the sunction dredging began once more, although the results were poorer still. The entire yield of twenty-eight days' work amounted to only 68 guilders. But this was to be the final year; although Maas and Dros had the right to continue dredging until 12 May 1892, they did not in fact resume operations. The 'lamentable era' – as Ter Meulen put it – lasting from 1886 to 1892 had come to an end.

18 Kinipple and Fletcher's British endeavour

In 1893 Ter Meulen took soundings above the *Lutine* once again. There proved to be twenty-three feet of water over the wreck, conditions favourable for operations with his own invention, the sand-diver. But there were further arguments for working on the wreck along the lines he had suggested. After J. H. L. van Deinse, at a gathering of the Netherlands Society of Mechanical and Nautical Engineers, on 13 February 1892, had given a lecture entitled 'On working on silted-up wrecks, with special reference to the *Lutine* and sand-diving', a committee was set up which brought out its report the following year. The committee consisted of the engineering experts H. Cop, D. S. van den Broecke and A. C. J. Vreedenberg, who in their report of 21 February 1893 stated their conviction that 'Mr Ter Meulen's plan provided evidence of prolonged and keen thinking and of great knowledge of affairs, and that it seems to them to be as good as it could be, in the absence of practical experience on various essential points.' The committee said that it regretted 'that owing to the shallow water that has persisted for almost twenty years (1867–85), Mr Ter Meulen has been prevented from obtaining any practical experience of working on the *Lutine*, and that when there was at last more depth of water, so that the site was at least navigable, his invention was actually ignored or, more precisely, was not seriously investigated or considered.' This was, then, a little moral backing, which Ter Meulen could very well do with, for 1893 was to begin with a great disappointment. On 23 December the enterprise formally cancelled the contract of 9 September 1873. 'This meant the end of my career as a contractor' was his sober comment at a later date.

The explanation of this rather surprising turn of events lay in the board's second decision. It had Lloyds sign a two-year contract with W. R. Kinipple, of Brighton. Johan J. Fletcher, a British engineer, had worked with success on several wrecks and was now put in charge of the future salvage attempts on the *Lutine*. Ter Meulen had come into contact with him early in 1893 and Fletcher did not seem disinclined to tackle the job together with the Dutchman. Ter Meulen thereupon drew up a scheme for their co-operation but met with his next disappointment when Fletcher chanced to remark that Ter Meulen would never succeed without a diving bell. Ter Meulen tried to persuade him otherwise but the Britisher would not give up his diving bell plan, since he was of the view that Taurel had in the past recovered the second half million with his *Hollandsche Duiker*. When Ter Meulen explained that nothing at all had been recovered with this apparatus, and that it had been the helmet divers who had brought everything to the surface, Fletcher still held firm to the diving bell idea since, by using one, they would be able to work under water with electric lights. Ter Meulen then told him that two divers who had worked for him in 1880–81, and recovered £12,000 from the wreck of the S.S. *Hansa* near Terschelling, had laughed at the idea of using electric light in the large space occupied by the wreck. One of them had said: 'When I first enter it of a morning, the water is fairly clear, but as soon as I have fetched a barrel (he had recovered 800 of them) the water is as thick as coffee and the strongest light would be of no help to me.' When Fletcher finally wrote that Ter Meulen had faith in his own 'baby', but that he, too, had a 'child', the Dutchman came back with the witty retort: 'Fine, but my baby is about thirty-three years old. How old is yours?' It may be that this answer convinced the Britisher better than any technical argument could. In any case, Fletcher came forward with a plan of real co-operation, but it meant that if he were to provide the capital for the steamboat and accessories and Ter Meulen the sand-diving machine, the latter would enjoy the profit over the first two months but would relinquish all further benefit and would have to hand over his sand-diver to the Britisher without more ado. Ter Meulen could not accept this as a basis for their co-operation, especially as even in a good season a month often went by without a single day bringing favourable conditions. Time had been passing while these negotiations went on, so that Ter Meulen could not possibly meet the enterprise's demand that on the basis of the capital collected he should give them the assurance that November that he would set to work on the wreck the following season. They added, as a threat, that in the case of a reply in the negative the board would entrust the work to the men from Britain.

On receiving this pressing letter, Ter Meulen did pay yet another visit to Lloyds' Dutch agent, as he had done in the days of the shell-fishers, but, as F. P. ter Meulen stated in his memoir, this agent continued 'to work against Uncle Willem, did not inform his own company of the situation, and insisted that the enterprise should sign a contract with the English company, firmly assured that the English would encounter no technical difficulties.' They had the capital, they had a steamboat, and being Englishmen they were naturally the Dutchmen's superiors ... thus went the argument more or less. And so, with the enterprise's blessing, Lloyds signed a contract with the new *Lutine* syndicate, with which the name of a certain Jaffrey was associated, besides those of Kinipple and Fletcher.

In the summer of 1894 Fletcher put in an appearance on Terschelling and was to return there many times, for he was in charge of the activities for five years. These activities, however, had miserably little result. In 1895, the first year it was possible to work on the wreck, their haul amounted to a value of 203 guilders (about £20), in 1896 to 55 guilders and in 1897 to approximately 128 guilders. During the period from 1894 to 1896 Burgomaster D. Reedeker of Terschelling received, as Receiver of Wrecks, the princely sum of three guilders and sixty-nine cents in the way of commission, while the board of the enterprise was able to distribute among the shareholders the sum of seven guilders and twenty-seven cents. Not a single gold or silver bar was recovered but merely coins, among them – as though to add insult to injury – scores of copper coins, whose value was put at nil.

In *De Ingenieur* of 2 November 1895 Ter Meulen reported on the results of the costly work done in the 1895 season. The British plan was to build an underwater wall round the wreck, remove the sand inside it by suction, and then to set two helmet divers to work. They made a beginning in September 1894. It was by then already too late in the season to accomplish much, nevertheless they did arrange 7,000 sandbags on the sea bed in a circle of approximately 250 feet diameter, though without being sure whether the wall contained the main part of the bullion. These sandbags remained fairly firmly in position during the winter, as the British divers discovered in the spring of 1895, although some parts of the wall had sunk into the sand, making it necessary to build them up again with 600 to 700 more sandbags.

Work was resumed in April. Now and again the British hired the two shell-dredgers which during the eighties had worked on the *Lutine* on their own account. They also equipped a small vessel themselves with steam engines, pumps and the like for applying suction and pressure to the water. Before this vessel could be used, however, it was necessary first of all for the shell-dredgers to clear the sand out of the area enclosed by the wall, and this they failed to achieve. The sandbags began to slip and spread about all over the enclosed area, and proved a great hindrance to dredging operations, the suction pipes jamming up against them.

In 1949 a seventy-five year old former employee of the firm of Maas living in Makkum could still clearly recall the proceedings during the four summers he had worked above the wreck. He recounted to a reporter how the British ordered a ship full of jute sacks from England, filled them with sand on Terschelling, and had them taken out to the wreck by fishing boat.

'The sandbags were thrown round the wreck in a big circle so as to form a dyke, and when the weather was calm and there was no current the English divers would go and look to see if the dyke was almost sealed. If they came across a hole they sent up a board with a number on it to indicate the number of sandbags that should be dropped overboard at that spot. They were tipped over the ship's rail in their thousands. So at last the dyke was sealed, working in this way, and we could begin to dredge again. But we dredged so hard that the dyke lost its foundations on the inner side and collapsed *on to* the *Lutine*. The bags got caught in the suction pipe and we had to *cut* them out. So we went on dredging until no more sandbags came up, but by then the pit we had dredged with so much difficulty had filled up again.

'It would have been one summer later that the English bought hundred and hundreds of oak piles from Friesland. Hendrik Stobbe, the smith, fixed steel spikes on the end of them, the divers drove them in one by one with a sledgehammer in a circle around the wreck, planks were placed between them, fastened by nails and heavy chains, and then we dredged again inside this wall. It was not long before this circular wall lost its foundations too, and shortly the whole sea was strewn with planks and oak piles ...'

When dredging inside the collapsed barricade had to be stopped, they began dredging outside it. Parts of the wreck appeared to be lying there too, which is not surprising since the investigations Taurel had made almost forty years before had shown that the *Lutine* site was fairly extensive. But success evaded them, even though 1895 was one of the most favourable years there had been for salvage operations. From April till September there had been sixty-one workdays when the sea was calm, after which there was only one good day in all the weeks up to 22 October. As a rule, the number of days on which they could work did not exceed thirty per year, but in 1895 there were at times five or six days in succession when the sea was as smooth as a mirror and they were sometimes even able to stay out over the wreck during the night.

In the meantime, the mood among the personnel was good, they worked industriously, and according to Ter Meulen the divers said: 'It is work for nothing, but we are well paid.' We have no information about the reactions among the contractors.[38]

Fletcher went on with his salvage attempts and the years passed.

In 1898 the Englishman arrived on Terschelling on 24 April, after fresh buoys had been laid above the wreck on 21 and 22 April. They set to work at once with suction dredgers, but strong north-easterly winds prevented regular work at a time when there was eighteen to nineteen feet of water at ebb tide over the wreck. 'A high ridge is approaching SE of the *Lutine*,' says a report, the last at our disposal. It seems that operations were brought to an end after this. When the syndicate was dissolved in 1900 there was hardly sufficient money left to wind up affairs, since all the funds it had been able to procure had been used to defray the cost of the operations. When, therefore, Binnendijk remarked that in view of these results it is 'rather surprising' that Lloyds should have signed a new agreement with Kinipple before the year 1900 was out, he was putting it politely.[39]

Ter Meulen, who was now completely ignored, spoke his mind. When he heard of the plans to build a wall and dredge out a pit inside it, he remarked mockingly: 'Now, that's not at all a bad idea, making a big hole above the *Lutine,* but they've forgotten one item – it needs a lid on it.' When despite good weather 1895 produced nothing, and the British went on with their fumbling, he said sarcastically: 'It's as if a priceless museum were being rehoused by a set of ham-fisted remover's men.' He did not confine himself, however, to critical comments but tried to make Lloyds see that Fletcher's method was not only pointless but actually disastrous, for it would be a long time before they could count on having such a depth of water above the wreck again as they had had in the years around 1895. He had Lloyds approached in London, first by a Dutch engineer living there and then by the secretary of the Netherlands Society of Mechanical and Nautical Engineers, who enclosed an English translation of his lectures on sand-diving. This, however, achieved nothing at all. At his wit's end, Ter Meulen had a card printed, which bore the legend: '1867–95. The sand-diver is the only man who can gather the scattered treasures of the *Lutine*', underneath which one could read: 'Dedicated to Lloyds by W. H. ter Meulen, Bodegraven.' This card was circulated, but although Ter Meulen received many approving responses, none were from Lloyds.

A step taken the following year, 1896, also proved fruitless. J. J. L. Bourdrez, a young engineer who was also a qualified teacher of English and could therefore express himself excellently in that language, was given full technical details of Ter Meulen's scheme and then — by way of introduction to a discussion he was due to have with them — he sent Lloyds an exposition, signed by no less than three leading engineers and chief inspectors of the Netherlands Rijkwaterstaat (department of bridges, canals, etc.) and by six professors at the Delft Institute of Technology, in which a plea was made for 'a fair trial on the *Lutine*'. When Bourdrez went to London in April he took with him details of a suitable scheme for financing the operations and a letter from Ter Meulen, in which the latter pointed out the inaccuracy of the view they appeared to hold in London that for years he had had every opportunity to test out his invention on the *Lutine*. But to quote his nephew and biographer once more: 'Lloyds closed their ears and ignored the matter.'

In 1898 Professor A. Huet published yet another recommendation for sand-diving in the magazine *Electra*, in which he termed it an invention of lasting value, and in 1899 Ter Meulen, then almost seventy, took up his pen once again, discussing in this same weekly the question whether sand-diving could be combined with sand-dredging. It all fell on deaf ears, and when in 1899 Ter Meulen also suffered a stroke of bad luck when the company he had recently set up (called, not without intention, the Hope Shell-Fishing and Salvaging Company Ltd) lost a very good chance of profit in a dredging job due to the action of two competing companies, his resilience was broken. He died on 17 June 1901. Ter Meulen should really be seen as yet another — belated — victim of the *Lutine*, and grouped with the hundreds who lost their lives on that evening of 7 October 1799. His will consisted of one small sheet, headed 'A sand-well'. It amounted to the ultimate defence of his invention, which had been often praised but never used.

19 England keeps the initiative

At the end of 1900 the contract with Kinipple and Fletcher's syndicate expired, yet December was not yet out when Walter Robert Kinipple Esq., of Devonshire House, 51 The Drive, Hove, Sussex, sent out a lengthy prospectus in which he spoke in optimistic tones of new salvage attempts in the coming years. Under the title *Lutine Salvage, 1901–02* he went to the trouble to give a brief item by item account of the history of the bullion ship and the salvage attempts made on her, following this by declaring that on the basis of his own and others' calculations there must originally have been at least £437,067 on board the *Lutine*, and further by explaining how it was that the syndicate had had so little success. He attributed this to the fact that no barrier had been erected around the site of the treasure to prevent the sand from creeping in, also to the fact that the operations had been carried out over too wide an area and too unsystematically. This go-ahead man of enterprise declared himself to be now so convinced that results could be expected for a moderate outlay that he made public a programme of operations and announced that he had already obtained a licence in his own name from Lloyds for 1901 and 1902.

After having let slip that in addition to much work and effort he had also devoted £1,750 to the enterprise during the past six years, he went on encouragingly, writing of himself in the third person: 'Having at last full and sole control of the operations, Mr Kinipple is now in a position to have his views carried out, more especially as he will have as his representative in charge of the salving operations at the wreck an old pupil of his, namely Mr John Scott, a Resident Engineer of the East Indian Railway, at present on furlough until October 1901.'

Kinipple seems to have been very impressed by this Mr John Scott. Among the factors which assured him that the prospect of success for the salvage operations for next year was much greater than it had ever been in the past was not only that a technician who had six years' experience of operations on the selfsame wreck would be in charge (this was evidently a reference to Fletcher) but also the fact that they would have the additional co-operation of the engineer from India. He introduced Scott as an experienced technician who had been in charge of large-scale works in India, from where he had just returned after building the largest bridge yet built in that country, for the construction of which it had been necessary to dig down through approximately 8,000 ft of sand and clay. What was more, this engineer was accustomed to carrying out orders 'literally and without alteration'.

A further promise of success lay in the fact that he, Kinipple, had provided £4,000 and that besides employing steam bucket dredgers and the usual divers' equipment with breathing tubes, they were planning to work with diving suits without such tubes (?) and also to make use of electric light. The fact that Ter Meulen's divers had declared that electric light was utterly useless was apparently a matter of no importance to Kinipple, though he may have been unaware of it.[40]

When the opportunity arose to become a shareholder in the New Lutine Company Limited at £5 a share, only £2,800 was subscribed. Nevertheless, Kinipple had his monopoly (for four years) and a beginning was made without delay. As early as 28 February 1901 John Scott was in Terschelling taking soundings above the *Lutine*. He had found seventeen feet of water above the wreck at high tide and eleven feet at low tide, which was two feet more than in 1899 and 1901.

Five months later, however, Van Buren Schele was obliged to report that on 2 August it had been decided, in consultation with Lloyds, their agent in the Netherlands and the directors of the company, 'to suspend operations for *this* year, since with only one sand dredger they would not be able to investigate the deep pit properly once it was ready. They considered this a wise measure and hoped to have two or three powerful sand dredgers at their disposal again the following year (Fletcher had used the shell-dredgers *Friesland* and *De Tijd*). Since 2 May they had spent about another 7,000 guilders.

The next year was little better, and for this the 'exceptionally unfavourable' weather conditions were partly responsible. Just as little was achieved in the succeeding years, operations being confined to taking soundings. All the same, Kinipple's licence was extended repeatedly, finally to December 1911. In November 1910 the New Lutine Company decided with the consent of Lloyds to employ the National Salvage Association to undertake the work on a trial basis. The new company, and Kinipple and Scott with it, thereupon disappeared from the scene and we hear no more about them. A new era dawned, an era that had at least one thing in common with the preceding one: nothing of any significance was recovered. On Terschelling this period is known as the '*Lyons* episode', and there are still many alive today who can tell about it with relish.

In 1910 the new British National Salvage Association Limited launched a prospectus with the aim of raising a capital of £79,800. It was emphasized that the company had been founded as a lifesaving and salvaging enterprise and that it should be understood that it would have nothing to do with what it called 'treasure hunting'. It is, in fact, true that with its vessel the S.S. *Lyons* the company did compete with the boats belonging to Zurmühlen and Dros on various salvaging jobs, although in an article in *Ons Zeewezen* (Our Shipping) in 1934, J. C. de Roever claimed that 'many doubted the serious intention of the salvage work'.

However this may be, the arrival of the S.S. *Lyons* in the harbour of Terschelling on 19 November 1910 'with an extensive crew in immaculate uniforms' attracted attention for more than one reason. The manager of the new enterprise was Charles A. P. Gardiner, who received £720 per annum for his services in addition to a further £6,800 'for the benefit of the licences he had obtained and for the information he was to supply', as Binnendijk records. Gardiner lived with his family on board the *Lyons,* which the people of Terschelling still speak of as being a fine ship. She was moored at the middle pier at West Terschelling, and on her side the word SALVAGE was painted in huge letters. As we said, the crew were dressed in uniform, which had never been the custom on salvage vessels, and true naval discipline reigned on board, which gave rise to a rumour that she was in part a naval vessel which in the past had been used for espionage by the British. It is improbable that this assertion (of which there may or may not be some echo in the remark made by De Roever, quoted above) was true; the British were certainly already in possession of good charts of the Terschelling area. More reliable information is known about the *Lyons* herself. Built at Glasgow in 1885, a steamship of 537 tons, she had done years of service on the channel crossings before setting out with the lighter *Bill O'Malley* for Terschelling. Before sailing she had been equipped with powerful suction apparatus for dredging sand and was completely fitted out for salvage work. As regards the work she and her crew performed there, we will quote from Binnendijk:

'On the night of 31 January 1911 the *Lyons* commenced sucking the sand from the *Lutine* site. She worked for eight hours, after which she had to stop owing to a N.W. wind coming up very suddenly. The first attempt was rather promising, two small shot about 1¼ inches in diameter, a few small pieces of wood and two small iron bolts being recovered. An exciting time followed, the *Lyons* working on the *Lutine* whenever weather permitted. Quite a quantity of material was recovered, such as copper sheet, copper nails, human bones, cannon, cannon balls, iron ballast, timber and many silver and gold coins. The two anchors, weighing 3,900 kilos, were also recovered.

'Great difficulties were experienced in those days with the burgomaster of Terschelling, who insisted that all articles recovered were to be handed to him in his capacity of Receiver of Wrecks. In the beginning the *Lyons* had to allow a policeman on board during the operations, but owing to the combined action of the Dutch and British authorities and the calls Mr Schröder made on the governor of the North of Holland and our Ministry at The Hague, the burgomaster's opposition was eventually overcome. At the Exhibition of Shipping at Amsterdam in 1913 the two anchors and 3 cannon were exhibited in the open. After the exhibition the Committee of Lloyd's contemplated having the two anchors forwarded to London as a monument behind the Royal Exchange, to replace the one of Sir Robert Peel. The plan was not carried out, but our office eventually received instructions to send the wooden stocks which bore the name of the *Lutine* to the Committee of Lloyd's, and this was done.'[41]

From Binnendijk's account we also learn that the licence under which the National Salvage Association was working was extended from time to time, and in March 1915 Lloyds had agreed to the Association's continuing operations until one year after the outbreak of the World War. But the *Lyons* was soon requisitioned by the British government and when the war eventually came to an end the company had abandoned the search for the *Lutine* treasure. An account in the *Lutine* archives of the articles recovered by the *Lyons* from 1911 to 1913 mentions: several gold and silver coins, 2 bower anchors, 15 cannon, iron ballast, copper, bronze, lead, etc., the sale of which by Alfred Schröder produced in all the sum of 1,377 guilders and 32 cents – approx. £135; the 'real salved value' after deduction of expenses being about 1,263 guilders – approx. £125. Of this the burgomaster of Terschelling, as Receiver of Wrecks, received 18 guilders and 94 cents, while the Dutch 'decretal salvors', as the inheritors of Pierre Eschauzier's estate were officially referred to, had to be satisfied with 37 guilders and 32 cents. All in all, a profoundly disappointing result, after three years' work on the wreck. The rôle played by the British in the salvage had come to an end for ever.

20 Van Hecking Colenbrander and Van der Wallen's coal grab

A hush had fallen over the *Lutine* after the *Lyons* had left, and the war had been over for two years when the landmarks on Vlieland and Terschelling, which had been removed by order of the Dutch navy, were put in position once more at the request of the *Lutine* committee. Soundings were also taken over the wreck in 1920. It appeared that five and a half fathoms of water were over the sea bed at low tide. But there the matter rested. Nor did the Delft students P. A. van Hecking Colenbrander (born in 1901, the son of a vice-admiral) and P. van der Wallen, who came from Brielle, hit immediately on the plan of testing out the dredging cage they had invented. Van Hecking Colenbrander, who was studying naval architecture, and Van der Wallen, who was studying electrical engineering, made their invention in 1923. It consisted of a cage made of wire netting with piping attached. Water was pumped into this pipe and came into contact with the sand below via upper and lower sprayers. The water from the upper sprayers passed through the wire netting, that from the lower sprayers travelled along the outside of the cage. Water would always be present above the wire netting cover, and therefore, according to a specification, 'it will find its way to the surface along the line of least resistance.' The cage had to be drawn up in order to bring the resultant slushy sand to the surface. Van Hecking Colenbrander seems to have been well informed about Ter Meulen's sand-drill.

However, it was not a 'dredging cage', with which experiments were made on the *Lutine* in 1924 and 1925. After 1923 the two students seem to have entered into partnership and by this time they wished to try out their invention on the bullion ship. The apparatus with which they went to work consisted, however, of an improved type of coal grab and a thick double spraying hose. Powerful jets of water were forced through these pipes and were fed into the grab as it was being sunk into the sea bed. On the outside, this grab looked rather like the familiar coal grab used in loading and unloading ships. The apparatus scraped or bit the sand out of the sea bed after it had been loosened by the jets of water. It would finally reach the bed of clay where the wreck lay and bring lumps of it to the surface, no divers therefore being needed.

After the two students from the Institute of Technology had talked over their plans with experts and made certain of some financial backing from their own circle, a small trial apparatus was constructed by the N. V. Haarlemsche Scheepsbouw Maatschappij (Haarlem Shipbuilding Co. Ltd). In Brielle, attempts were made to assemble a demonstration unit, but it was soon clear that this unit was of insufficient capacity to feed the apparatus as required. The two young operators, who seem to have had good connections, then enlisted the help of the fire brigade from The Hague and the military at Waalsdorp camp nearby. A derrick was erected in the camp grounds, and with the help of the army's fire-hose they got the grab into the bed of a pool of water there at the rate of approximately $3\frac{1}{4}$ feet a minute. This was in 1924.

Contact was then sought with Lloyds, and a small coal grab was bought and modified until it seemed suitable. The young men went off with this piece of equipment, to Den Helder, in the north of Holland, and there it was tried out in the harbour with the help of a tugboat belonging to the Wijsmüller Bureau. However the presence of a great deal of sludge hampered the work, and so a second trial was held on the *Alechia*, which had been wrecked off the coast of Texel. The Texel firm of Dros, in charge of the salvage work, lent its full co-operation. There were winches and pumps on board and even the wrecked ship's boiler proved of use. This trial was more successful, but it was still impossible to say with certainty what capacity was desirable for operating with success in the Schellinger shallows where the *Lutin*e lay. Meanwhile, time was pressing. A syndicate was formed under the directorship of Jonkheer J. O. de Jong van Beek en Donk, ex-governor of Curaçao, to raise the sum needed for carrying out the salvage agreement which in the interim Van Hecking Colenbrander had concluded with Lloyds. Signatures were placed on 21 June 1924 and it was specified that the syndicate would lapse on 31 December of that year.

Although apparently the two inventors had not quite finished their preliminary work, the contract with Lloyds required the work to be started at short notice. The Katwijk steam lugger *Gerard* was chartered from W. Nijgh in IJmuiden, and pumps, winch and hoisting-tackle were erected on the vessel without delay. *Gerard* set course for Den Helder during the night of 2 July. Here all the equipment was made ready for an experiment close to the Onrust shoal, in the Marsdiep, on 8 July. Yet further alterations proved necessary. The next day all the equipment was tested again and luck was with them, for the grab turned out to have been let down over a piece of wreckage which was brought to the surface at once from a depth of about fifteen feet under the sand. On both the following days they managed to make the grab penetrate the sand to a depth of almost forty feet within half an hour.

On the 16th an attempt was made to determine the site of the wreck with the help of a pilot from Terschelling, but this was prevented by fog. On 17 July the *Gerard* steamed out to sea again, but although experts had affirmed that in April the water was still twenty-four feet deep over the wreck, it now proved to be a mere six feet. The *Gerard* drew twelve feet and could not be used. A dinghy laid a buoy over the wreck, another demonstration following in the Schuitengat for the benefit of Lloyds' representative Wilkins, and then they had to wait for the tug and salvage boat *Stortemelk*, belonging to the Terschelling shipping company of Doeksen, which was to replace the *Gerard*, but which could not leave the shipyard in Harlingen until 4 August.

The apparatus was set up on 6 August and on the 7th it was put to the test over the *Lutine*. Although the grab could not yet be let down to the depth required because of a layer of shells, a few lumps of cement were brought up, coming from a block which the *Lyons* had had sunk there in 1912. Two days of bad weather followed, but it was possible to resume work on the 10th. Unfortunately, the pressure pipes then burst and the *Stortemelk* had to return to harbour. A fresh start with new pressure pipes was made on the 12th, but they did not get down as far as the bed of clay. And so began a series of operations which had very little result and during which they lost the entire equipment on 28 August. With the help of a new sprayer they managed to recover the apparatus and pipes from a depth of about twenty-three feet under the sand within an hour, and without any damage having been done.

Bad weather began to play tricks on them as early as September: on the 19th an anchor chain broke and 2 October saw the last operations in 1924. The wreck had yielded up a few pieces of wood, lumps of rust and some cement. In one of the lumps of rust a copper pin and an eyelet were embedded. Trips had been made to the wreck on seventeen days after 6 August, only eight of which had been suitable for regular work. In January 1925 Van Hecking Colenbrander sent out a circular describing a syndicate for recovering valuables from the *Lutine*, in which financial support was requested for a new salvage attempt planned for April and requiring at least 50,000 guilders. Lack of the requisite funds forced the student of naval architecture to not take up Lloyds' offer to prolong his contract until 31 December 1926.

The first trip out to the wreck in 1927 did not take place until 10 June. But the Netherlands had another of its bad summers and work could only be carried out over the wreck on a few days in August and September, when trials were made on 14 August under the supervision of an engineer from a factory which had made its material (called 'explosive air') and the associated equipment available. The efficacy and handling of this explosive in the sand, at a depth of over fifty feet under water, proved satisfactory. An exceptionally stormy October followed, so that only two trips were made to the site, and these were fruitless. It was not until the 30th that the exact location could be determined again, after which the suction dredger *Volharding* (Perseverance) appeared on the scene on 1 November, to try to suck the sand away so that the explosive and grab might have easier access. The trial with the suction dredger was quite satisfactory, but on 2 November the weather changed once again and this signified the end of the operations for that year and for the Van Hecking Colenbrander and Van der Wallen period.

Is it possible that the facts concerning the wreck and the many salvage attempts on her were not very well known outside the Netherlands? At all events, a certain Maximilian Neugebauer, of Vienna, wrote to Lloyds in July 1926 that he was coming to Terschelling at the beginning of August with machines and personnel to try to salvage the *Lutine*. Apparently this was to be taken literally, for while this enterprising Viennese wrote that he had already made an attempt in 1911 'with inadequate means', which failed because of the tremendous current, he now planned to tow the wreck to the coast with the aid of mammoth pumps and air-drums, as if the ship were still all in one piece. He generously offered Lloyds 50 per cent of the treasure if he succeeded, but when Lloyds replied that a contract could only be signed after more information had been supplied and after the Dutch parties involved had been consulted, nothing further was heard from him. Nothing is known, either, of his alleged salvage attempts in 1911. He was not, however, the last foreigner to show an interest in the enticing bullion ship.

21 Well-known salvors take the plunge

Two serious salvage firms, the N. V. Stoomboot Maatschappij Texel (Texel Steamship Co. Ltd) on Texel (Directors A. Dros and P. Dros) and the N. V. Scheepvaart Maatschappij G. Doeksen en Zonen (G. Doeksen & Sons Shipping Co. Ltd) of Terschelling, launched a prospectus in August 1928. The latter of the two companies described its activities, not without pride, as 'salvaging, shell, shingle and sand dredging, diving operations and clearing away wrecks by means of explosives.' They had carried out the 'salvaging of the stern of the *Liberty Glo*, an American vessel which was flung high up on the beach of Ameland with her back broken during a howling gale at the end of 1919. The work of salvage went on for three raw winter months, violent storms time and again destroying all that had been achieved, yet in the end the suction-dredgers managed to make a channel hundreds of metres long in the breakers, through which the ship was hauled back into the sea.'[42]

It was really not surprising that Doeksen and Dros, who were more at home than anyone else in the waters and among the sandbanks of the Wadden Islands, should have wanted in the long run to have a try themselves. Anyone reading their brochure must certainly have been convinced that this was no gamble and that the salvors were thoroughly aware that it would be no easy task. One must not, they warned, underestimate the enormous difficulties involved in these exceptional salvage operations, but must make special allowances for the fact that the *Lutine* lies in a very unfavourable and dangerous spot in the shallows, where one can only work when the weather is good.

However, they were full of hope. They intended to suck away the sand on and around the wreck, to blow up the lumps of rust and pieces of wreckage and clear them away with the help of divers, after which the site would be well worked over by grabs. The divers would be provided with what were then the most up-to-date facilities, such as electric light (for the umpteenth time!), telephone, etc. Doeksen and Dros were not financial visionaries but quiet, reliable, hard-headed northerners. They frankly informed people who might be interested that the work was estimated to take three years and that an average of 100,000 guilders would be required per season, 300,000 guilders in all, for 'many difficult and dangerous operations calling for dogged powers of endurance'. The job of sucking away some hundreds of thousands of cubic yards of sand would alone keep the suction-dredgers continually occupied. According to what they said, the men from Terschelling and Texel had planned at first to finance the work entirely on their own, but later they defrayed half the cost themselves and offered shares for subscription by others for the second 150,000 guilders.

This was in August 1928. The *Lutine* land beacons on Vlieland (on the outer line of dunes) and Terschelling (on the Noordsvaarder sandbank) were set up once more and soundings showed that the depth of water over the *Lutine* had increased considerably during the last few years, justifying the expectation that 'a period of conditions suitable for salvage work has set in', as the prospectus accompanying the brochure concludes. Whereas about a fathom of water had been sounded in 1926, in May 1928 it was established that there was well over three fathoms, which was far more favourable. After they had waited – apparently for reactions from parties interested in supplying the necessary finance – at the end of October 'The Society Incorporated by Lloyds Act 1871 under the name Lloyds' and the two salvage companies, the Texel and Doeksen & Sons, signed an agreement whereby the London organization granted the contracting parties the right 'to carry out salvage operations' for five years on the familiar basis of 'no cure, no pay'. Lloyds was no longer as demanding as it had been in the nineteenth century. It agreed that the contractors should receive 70 per cent of the profits and contented itself with 30 per cent, out of which the so-called 'decretal salvors' (the rightful claimants by virtue of a royal decree of 14 September 1821) would receive another 10 per cent. The contract ran until 1 November 1933, but was prolonged on that date.

Even before the agreement had been signed, the question again arose as to who really had the right to the treasure. It seems that it was the burgomaster of Terschelling who was now hoping for a windfall. We have had sight of a letter from Alfred Schröder, Lloyds' agent in Amsterdam, in which on 19 June 1928 he informed the burgomaster-cum-Receiver of Wrecks of the island: 'Your writing that the Chief Receiver of Wrecks was the founder of the Dutch *Lutine* enterprise according to the royal decree of 14 September 1821, no. 65, may be very true, but in my opinion there can be no question of any Burgomaster-Receiver of Wrecks deriving any right from this. The function of Chief Receiver of Wrecks happens to have been abolished by the royal decree of 23 August 1852, and besides, it is an open question whether the rights granted to the Chief Receiver of Wrecks, Eschauzier, at that time are not to be regarded as personal rights. Discussions about this went on all through the nineteenth century, and as soon as there was any question of actual salvage work being carried out, these matters were settled on the spot by a friendly agreement with the burgomaster-wreckmaster.' After receiving this letter the burgomaster no doubt would have taken the line of least resistance, although it is undeniable that in the royal decree there is no reference to P. Eschauzier personally, but of the 'Chief Receiver of Wrecks of Terschelling', although this general definition is linked with Eschauzier in the same sentence, since it mentions 'the plan put forward by him in his letter to the Governor of North Holland dated 15 June last'. It is difficult to maintain that chance successors of P. Eschauzier, who neither made any plans whatsoever nor invested a penny, could claim half, or a portion, of what was recovered. As Schröder observed, on previous occasions the question raised by the burgomaster of Terschelling had indeed been settled to the satisfaction of all concerned.

In the meantime, the island official, like so many before him, had been haggling about the skin of a bear which was far from shot. Notwithstanding the large-scale modern equipment and the fact that the contractors did all they could to make a success of their plans, it soon had to be admitted that local enterprise was also unsuccessful. The tormentress who in the past had shown little sympathy for Britons did not allow herself to be outwitted by Dutchmen from the neighbouring islands.

Two reports about the negative results are at our disposal. The first comes from N. S. Binnendijk, the associate of Lloyds' agent in the Netherlands, who reported in *The Log*: 'During the first years the salvors were rather unfortunate. Apart from the fact that there were no long spells of good weather, the suction pipes broke occasionally owing to the swell. Furthermore, the hopper barges used (the sand and shells sucked up were discharged into these barges to prevent the sand running back into the pit) did not work satisfactorily.' In the summer of 1933, after five years' effort, the contractors themselves declared (in a brochure announcing another new scheme) that for some time they had been trying to clear the *Lutine* of sand by means of suction-dredgers. 'We have not been successful to date. Our dredgers have, however, sucked up various fragments of the *Lutine* ... If there should be an increase in the depth of water at the site of the *Lutine*, which, judging by past experience, is not unlikely (the current is always bringing about changes there), the suction-dredging could be carried on with far less difficulty. In view of the scheme outlined hereafter we have suspended suction-dredging for the time being.'

One needs to be fully aware of the determination, competence, and long years of maritime experience of these salvors and their men to realize that the statement 'we have called a halt' – and not after a year or two, but only when five years had passed – was not made readily, even though new prospects were held out. Dros and Doeksen's enterprise did not miscarry because of hare-brained ideas and risky plans, but quite simply because of what sociologists tend to call 'the concrete situation'. The submerged millions proved to be unattainable by ordinary methods, and it was now the turn of extraordinary and as yet untested ones. One such method had already been tried out in the meantime.

In 1931 the contractors had got in touch with Mr Victor Hugo Duras, from Washington D.C., a Supreme Court attorney. He represented a group which had been successful in salvaging sunken treasures in Panama using a 'patented radio metal and mineral finding machine' which could indicate where concealed gold and silver lay. When tests had been carried out on the *Lutine* site the American had to admit that the machine had insufficient capacity in this instance. The dredgers *Texel* and *Volharding* simply continued as before, the latter dredging, the former working over the ground it had covered, yet neither the four divers, among whom were Sperling and Bijlsma, nor the special polyp grab, which was mounted on the *Texel* and designed to grab within the wreck, were able to recover much of value. A handful of Spanish piastres, a few cannon balls and a single gold coin were all. And then it was 1933 and Beckers' year.

22 Beckers and his tower

The graveyard at West Terschelling, near the Brandaris lighthouse, is mainly a seamen's churchyard, as is indicated by many stones on which sailing vessels and steamships have been chiselled – a tradition continued well into this century. Here anyone who has taken an interest in the story of the *Lutine* will come across many familiar names. The Eschauziers have preserved the memory of their forefather Pierre, founder of the Lutine enterprise, with a freestone monument which adorns the family grave. Reading the inscriptions – about pilots drowned on duty, captains who commanded famous ships, brave rescuers who gave their lives for others – the visitor is suddenly confronted with a stone whose inscription is striking yet which makes him wonder whether later generations will understand what lay behind its doleful message. This is the tombstone of Jan de Beer, born 6 October 1893, died 5 June 1935, who 'suddenly lost his life so tragically while working as a diver in the Lutine tower. Deeply mourned by his wife Trijntje Kooiman and child.'

On Terschelling people will long remember the episode in the history of the salvage attempts in which the *Lutine* tower, also known as Beckers' tower, played a part. But will people outside the island still know in years to come what sort of structure it was in which young Jan de Beer lost his life? A tower *of* the *Lutine*? A tower *on* the *Lutine*? A *tower* in the sea?

The year 1933 was not a week old when the Dutch press reported spectacular new plans for bringing the *Lutine* gold to the surface. They came from Frans Beckers, of Gennep, and were in earnest. A new invention had come into being and steps were quickly taken to try it out. Doeksen and Dros entered into a contract with the designer, an industrialist who had heard of the *Lutine* during a business trip to Terschelling, and on 4 May 1933 they addressed themselves to the public once again with another brochure and a prospectus. They recalled their own abortive salvage attempts – 'large sums of money were expended but it was not possible to clear the sand away permanently' –

and gave a detailed account of the essential features of the Beckers method, the installations for which had already arrived on Terschelling. They described the method as one which made the salvage work far more independent of silting-up and weather conditions, and went on to report that 'Mr Beckers proposes, in collaboration with ourselves, to place a steel tower over the wreck, ten feet in diameter at the top and forty-two feet at the base. We need a few consecutive days of good weather for its installation. As the tower, which is open at both ends, is lowered into position, the water and sand will be pumped out by a powerful motor pump. Once it is in position over the wreck, on the spot where the masses of rust lie, the upper part will project some metres above the surface and below a closed-off space will have been created in which there will be access to the wreck in all normal weather conditions and the work of clearing away can then proceed apace. The tower is equipped with a lift, lighting and ventilation installations. The power units supplying these facilities are mounted on a platform fixed to the upper part of the tower. Underneath, in the interior of the tower, there is about 110 square metres of working space. A number of workmen can be put to work here and everything that has to be cleared away so that the gold and silver can be reached can be brought up by the lift. The pumping equipment has been supplied by N. V. Van Zaal's Machinefabriek in Maassluis (Van Zaal's Engineering Works Ltd), the lift, etc., by Stork Hijsch (Stork Lifts) in Haarlem...All the preparations will be made with the greatest care and precision and we shall do all in our power to ensure that in the coming season the salvage work will be successful.' It was explained that on the strength of the contract with Lloyds, Doeksen and Dros would receive 70 per cent of the proceeds, Lloyds 30 per cent and Frans Beckers, as designer of the salvage equipment, 28 per cent of the 70 per cent. Fourteen per cent would be handed over to the shareholders, so that 28 per cent would be left for the salvage companies. The latter proceeded to issue a limited number of shares, each giving the right to 1/20,000th part of the gross proceeds of the entire *Lutine* treasure, which, it was pointed out, would amount to at least 900 guilders per share, if the total sum spoken of in the royal decree of 1821 were to be recovered.

Even before the Kas-Vereniging N. V. (Cash Office Association Ltd) in Amsterdam had received the subscription certificates, the inhabitants of Terschelling were able to see for themselves that the invention of Beckers, malt manufacturer from Brabant in the south,

Van Hecking Colenbrander, marine engineering student at Delft, with the 'improved coal-grab' he and van der Wallen invented, on board the *Stortemelk*, late summer 1924.

The *Texel*, belonging to the firm of Doeksen, in action over the *Lutine* (1928–33).

The polyp-grab on board the *Texel* (1928–33). It brought a good deal to the surface, but no bullion.

The raising of a cannon was always an event which appealed to the imagination (1928–33).

A *Lutine* still life: a small part of what was recovered during 1928–33. Centre, a wooden block with metal sheave.

The first of Beckers' towers in the inner harbour at Terschelling (from a postcard).

The same tower after being seriously damaged on the night of 25–26 August 1933.

The second tower after arrival at Terschelling on 24 May 1934. Right, the cylinder that would be erected on the conical base.

The second tower in the breakers over the Shallows, autumn 1934. Two steamers in the background.

Some of the coins recovered about 1930. From left, a copper Catherine II five kopec piece (1793) and copper George III halfpennies.

The ship's bell from the *Lutine*, recovered 17 July 1858, first hung in the Royal Exchange in London and then transferred in 1957 to the Underwriting Room in the new Lloyd's building in Lime Street.

ex-engineer in the German Navy, Dutch citizen, a small and active man in his sixties but ever cheerful and ready for the fray, as the newspapers sketched him, was to be taken seriously. As early as 21 February 1933 a beginning was made in the shipyard at West Terschelling with the discharging of material for the projected structure. In April the tugboat *Texel* took a complete pumping and suction unit on board and towards the end of the month boring went on on a large scale. On one day no less than 102 borings were made, which brought to light that the wreck of the *Lutine* was lying about fourteen yards down under the sand and that the depth of water over the sea bed varied from approximately nine to sixteen feet with the ebb or flood tides. Moreover, it was ascertained with the suction pipe that the entire wreck was lying below a layer of clay between five and ten feet thick. Its length was found to be approximately 62 feet, its width approximately 33 feet. The layer of ammunition measured about 40 feet by 33 feet and the mass of rust into which this layer had been transformed was presumably about 15 feet thick.

In the course of May 1933 the pattern of the preparations began to emerge. News reports from Terschelling kept pace with the developments:

'May 27th. The tower is ready. During the trial run, the pumpbarrel has shown itself to be in order. A heavy Jupiter engine will soon see to it that all the sand and water inside the tower will be emptied out in no time.'

'June 13th. The tower which will be placed over the wreck has been shifted from the inner to the outer harbour and is now lying ready to be towed outside in the next spell of fine weather. The salvage company is in direct communication with De Bilt (meteorological office) and will be notified when a spell of several days of fine weather with a southeasterly wind may be counted on. The plunge will only be taken when the weather is fine and the sea calm.'

There was *one* man in the Netherlands who did not place so much confidence in the tower, and even issued a warning not to take the intended plunge. This was Professor J. A. van der Kloes, of Delft, who argued in a technical journal against the placing of the tower on the wreck. 'It looks quite sound, yet I foresee failure, a wholesale collapse. Sagging to one side, toppling over, etcetera, etcetera. It is to be hoped that they will leave off pumping in good time.' But the papers continued to make encouraging and almost triumphant announcements: 'July 21st. Tower taken outside the harbour at 10 o'clock. At 8.41 in the evening they started lowering the colossus and at 9.15 a.m. the sea bed was reached.' A quarter of it still projected above the surface of the water, but we read 'the tower is slowly sinking down vertically into the sand'. J. van Drimmelen, a diver who had worked on the wreck many times before, had also appeared on the scene by then.

July 22nd. Towards ten o'clock in the evening, after some technical hitches (the suction pipe of the *Neptunus* became choked,) under the threat of deteriorating weather conditions (which afterwards turned out better than expected), and with the co-operation of the tugs *Texel* and *Volharding*, which started dredging, 'the strangest tower these waters ever saw' was placed in position. 'The tower is standing and this means that a workshop has been built above the treasure', it was stated with satisfaction.

'July 24th. Yesterday, Sunday, the lower section of the container, in which the powerful engine for the suction plant has been built, and which weighs approximately 35,000 pounds was slung up on the large derrick, which has become available now the cone has been taken out to sea, so that when the wind is fair they can put out to sea at once to continue the work. The salvage tower now stands bolt upright over the wreck.'

'July 26th. Yesterday the tug *Holland* towed the barges carrying the cylinder and engines out of the harbour. This morning the cylinder and engines are to be installed in the tower over the wreck. The weather is good.'

Meanwhile, Martin Sperling, referred to as 'the well-known diver', went down in the salvage vessel and examined the situation in the interior of the cone. His findings seemed favourable. Sperling went down in a diving suit weighted with sheets of lead on chest and back, a 2,000 candlepower electric lamp being lowered down to him. His first find was a long pipe which had been inserted into the mass of rust on a former occasion, by way of fixing the position, but which had snapped off later. As for the other two pipes, one proved to be standing quite clear, but the supports of the other were bent. What's more, the diver also came across a hole about thirty inches deep between the walls of the cone, making him conclude that a breach from the outside must have occurred during the dredging after all. In spite of this discovery, operations continued. On 14 August a start was made with pumping interrupted by diving investigators. For these the German diver August Istemaas, who had been a professional for nineteen years, used a new method of diving without air hoses from the surface, using a calcium regenerating cartridge to breathe through.

Meanwhile, the cylinder still had to be placed on the tower. On 17 August the shell-dredger *Neptunus* sailed to the *Lutine* for this purpose, but the strong west wind prevented the plans from being carried out, although the great suction pipe was lengthened to such an extent that it was possible to reach the mass of rust 'under which the gold lies'. The ship put back into port and waited for better weather. When that came on Friday 25 August, a glorious day, the *Neptunus* sailed again. On Saturday the 26th the future looked very bright. But when the *Neptunus* neared the tower it was clear that something had happened. The colossus was leaning to one side, heeling over towards the S.W. It had clearly been subject to destructive forces.

The divers Sperling and Istemaas went down at once and the devastation they found there made the German say to Beckers, 'pale with emotion', 'Der Turn ist gesprengt!'(The tower has been blown up!) The scene underwater was indeed rather disastrous. Just above the sand and about twenty feet under water, two large plates, each over three square yards in area, had completely vanished from the conical section. This destruction, so the divers imagined, (and Beckers agreed with them) could not possibly have been caused by the sea, as the hole had appeared in the lee side (where the waves did not beat against the tower wall). Foul play was immediately suspected: 'small cartridges of dynamite were probably fixed to the outside, under water, and the violent explosion drove the plates inwards.' It seemed even more feasible that an explosion had taken place when a jetsam collector from Terschelling reported that he had heard an explosion at sea about three a.m. in the morning of Saturday 26 August.

Who could have played such a malicious trick? The attack must have been carried out in a choppy sea, for the salvage vessels had been at the tower all the time when the weather was calm. It caused a sensation, the severe blow suffered by the shipowners being universally deplored. Their work for 1933 had ended as far as the Beckers method was concerned, and so on the 28th the *Volharding* and the *Neptunus* had to steam to the scene of the wreck with derrick and hoisting tackle. They raised the ravaged tower without much difficulty. The same day it arrived back in the harbour, where the police impounded it and the public prosecutor held an enquiry while the public's imagination began to ferment. In addition to the stories about the jetsam collector who had heard an explosion, there were rumours about a threatening letter which Beckers was said to have received some time previously and to which he had paid little attention at the time. Terschelling was a-buzz with rumours.

However, during the evening of 29 August it was announced from the island that: 'This afternoon the firm of Doeksen made a thorough examination of the damage to the salvage tower. Four plates proved to have been stove in on one side and all the bolts have worked loose. We are now convinced that this damage has not been caused by an explosion, as the divers' reports had it.' At this juncture differences of opinion arose between Beckers, who maintained that some foul play had certainly occurred and had unfair competition in mind, and the salvors, who reached quite a different conclusion. They declared that Beckers' tower was destroyed by the weight of water and sand. Their argument went as follows: 'The object is constructed in the shape of a pure cone of armour-plating 16mm thick of Krupp manufacture. As Beckers planned to have the work done in dry conditions in the cone, he had placed a packing of grease between the plates to render the wall of the cone watertight. Under the influence of the salt water, this packing began to work dangerously loose and on 29 August it was hanging in long streamer along the outside. The consequence was that play developed between the plates, the cone losing its rigidity.'

Doeksen and Dros were soon proved to be right. Professor van der Kloes, who had issued a warning earlier on, gave a detailed account of the probable course of events in the journal *O.T.A.R.* (Sept., Oct. 1933). The professor declared that the difference in water level inside and outside the cone could not but cause quicksand to form. This was not spread evenly round the base of the cone, but formed at one point along the circumference, determined by the suction tube and irregularities in the sea bed. It started with a small hole under the rim which became larger and larger and was first noticed when it was already thirty inches across. In proportion as the upward surge of the quicksand increased in volume and strength, an increasingly larger part of the inner wall of the cone loosened, until it was finally crushed inwards by the pressure of the sand and water outside. This would have been accompanied by an explosion, in Professor van der Kloes's opinion, for between 200 and 300 bolts had, after all, 'detonated' simultaneously. 'If one also bears in mind that over and over again the operators have been forced by wind and weather to leave the tower in a rough sea for considerable periods of time, exposed to the play of the elements, it is no wonder that the tin lampshade should have given way in the end.' The result of foul play? 'I've no objection to them trying to delude themselves and others, so long as they don't come forward with a repaired tower or a new one in the same form,' he wrote, showing scant respect for Beckers' invention.

Beckers and the salvors, however, did not leave it at that. We do not know whether a letter dated 14 September 1933, from the Dutch consul-general in Paris to the burgomaster of Terschelling, had any result. In it the writer asked for the name of the 'owner-contractor' of the damaged tower, as there was someone in Paris who 'wished to get in touch with the contractor in connection with the invention of a new salvaging apparatus'. It was true, though, that about ten days after the tower had been put out of action the *Volharding* and the *Neptunus* were once more busy suction-dredging by means of suction pipes twenty-four and twenty inches in diameter, while on 19 September Dros and Doeksen announced in a circular that they and Beckers would carry on.

In this circular they informed the shareholders that should the proposed working method (after sufficient sand had been sucked away) of breaking up the lumps of rust with explosives and bringing the gold to the surface with the help of polyp grabs prove unsuccessful, operations would be carried out with a new apparatus. As far as the financial side was concerned, Beckers had shouldered the cost of his tower unaided, launching a small company, known as N. V. Lunex, Maatschappij voor Bergingswerken (Lunex Salvage Co. Ltd) for the purpose. The setback – which he continued for a long time to ascribe to the work of ill-wishers – did not discourage him in the least. The optimistic malt manufacturer had a new and stronger tower constructed and in the meantime the salvors tried working with their suction-dredgers. On 10 November reports stated that work was still being carried on energetically. In that month the five-year contract with Lloyds expired and was prolonged for another year, while the salvors made new and far-reaching plans with their 'sub-contractor' for the spring of 1934. Beckers remained in the news and the coming of spring was eagerly awaited.

In the winter of 1933–34 a second tower approximately sixty-five feet high was built by the firm of H. Jonker & Son on Bickers Island in Amsterdam. It consisted of two parts: a cone-shaped lower section ('lampshade', to use Professor van der Kloes's term) and a cylindrical upper section. The lower section of this tower, which was welded in the normal way, had a diameter of about forty feet at the bottom and thirteen feet at the top, and weighed thirty-five tons. On 16 May both sections were hoisted on to a boat by means of a floating derrick, after which the shipment left on the 23rd. At nine o'clock in the morning of 24 May the new tower arrived in the harbour of Terschelling, but it was the middle of June before the smaller upper section, which weighed roughly as much as the conical base, could be fitted on to it by a derrick. Yet in August the tower was still lying at the quayside. Newspaper reports stated on the 8th that Beckers had been instructed to place his tower over the wreck immediately. But by 24 August this had not yet been done and the shareholders became impatient. Mr H. S. Goldschmidt wanted to form a committee to institute legal proceedings to reclaim the subscription money, but the next day the newspapers reported Doeksen's reply: 'We are forced to wait for fine weather, the sea is still too choppy.' However, on 1 September the tower was successfully placed in position over the wreck and after some reverses on the 12th a start was made on the 17th with dredging it clear of sand. Rather on the late side, one will say, but one should not forget that dredging operations had been going on normally in the meantime. Whenever it was *Lutine* weather, i.e. easterly wind or a slight breeze from the south, both suction-dredgers steamed out to the wreck and from time to time the divers went down. What did they bring up?

A newspaper report of 28 June 1934 gave some idea: 'The divers have laid hands on a great many cannonballs and heavy timbers and hoisted them to the surface. In this manner the wreck, which has first of all been cleared of sand, is completely broken up. Among the items hauled up are cannon breech-blocks bearing the date 1796. The cannonballs form a solid mass of rust, which has first to be removed before the place in which the gold must be can be reached.' July was marked by fine weather and this was very beneficial for the salvage work. On the 26th the *Texel* and the *Volharding* came in again with a cargo of wreckage. A correspondent wrote: 'The warehouse is gradually becoming full up. One finds great masses of heavy oak beams there, pieces of the keel, thick knee timbers. All this wood is still as sound as a bell and hard as metal. A pile of iron ballast, a mountain of cannonballs four and six inches in diameter and a large cannon have been placed near the entrance. Then there is copper sheathing, copper plates and so on.'

Although all this material was of little value, the breaking up of a wreck which had such an illustrious name attracted much attention. Reporters also flocked to Terschelling from outside the Netherlands, though they were not all as active as the French journalist Maurice Broc. He was determined to see the wreck with his own eyes and approached the Doeksens, who had no wish to deprive him of the sensation. He got into a diving suit which seemed terribly heavy to him – each boot weighed over ten pounds, over twenty pounds of lead was on his chest and the same burden on his back – and was let down. He reported on what he had seen on that 14 July 1934 in the Paris newspaper *l'Intransigeant* although the following text is based on the version published in the Dutch paper *Nieuwe Rotterdamsche Courant*:

'The chiaroscuro on the bottom of the sea puts him in mind of a cathedral. The actual discomfort is then over; the first moments in the water were often difficult. The breathing quickens involuntarily. By degrees forms take on an outline. He and his companion are in a pit about thirty-three feet across which they had excavated in twenty-five feet of sand in order to get down to the wreck. At the bottom of it utter chaos reigns. Though still partly buried under the sand and studded with shells, the wreckage reaches out to embrace him, as it were. The suction and the gentle swell force the men to go on hands and feet. With the help of crowbars and levers, his companion prises large pieces loose and he helps him to wind chains round them, which demands a gigantic effort. Broc sees this man's lips moving through the glass in front of his eyes, which serves as a porthole. When the crane raises a heavy item the water becomes clouded with churned-up sand. In front of him, a gap yawns open in the side of the ship and a cannon stands next to piles of cannonballs. Half an hour later they are on the surface again. It was the first cannon to be brought to light from the *Lutine* that day, and according to his own account, the boatman credited this stroke of good fortune to the presence of the French journalist.'

Let us now return to Beckers' tower. Brought into use in mid-September, it was very soon severely tested. A violent storm sprang up, which ruled out all work above the wreck. But when the *Texel* and the *Volharding* arrived at the tower on 30 September, it proved to have held magnificently and to have incurred no damage at all. It proved possible to suck up the remaining sand in the tower so that the 'indoor' salvage operation started the next day. At the time the tower was standing with its base sixty-five feet below the surface of the sea and had reached the solid layer of clay, covering about 120 square yards of it. But the hopeful divers had only just started working under the new conditions when the weather worsened again, and an enforced break lasting three weeks followed.

Work was resumed on 24 October, when it was observed with satisfaction that the tower and machinery had not incurred any damage at all. No salvage of any importance was noted, however. On 21 November the Dros-Doeksen contract with Lloyds was prolonged for a year and at the same time it was announced that work would cease during the winter months. The operators, who had recovered only one coin, nevertheless termed the results 'not insignificant'. The tens of thousands of cubic feet of sand lying on the wreck had been removed, the wreck itself had been almost completely cleared, and plans were made to explore the *Lutine* territory 'foot by foot' in the following spring.

The *Lutine* was left in peace until the beginning of May 1935. The *Texel* and the *Volharding* took the lead and on 10 May work on the tower was resumed. But success evaded them this time too. The weather was very bad that summer and there were only twenty-seven days between April and September on which it was possible to work on the wreck. June 6th was a day of ill-omen on which the *Lutine* claimed yet another victim; the first for 136 years. This was the diver, J. de Beer, who was working in the tower with his colleague E. Mettinga. A third diver, J. Bijlsma, was standing at the signal line and at a certain moment did not receive any answer to his signals to De Beer. He immediately signalled Mettinga, who surfaced without delay, descending again at top speed on learning that De Beer did not reply. He found his colleague lying motionless at the bottom and brought him to the surface, but the man was already dead. 'The cause of death has not yet been established', ran a press report.

This accident cast a shadow over the salvage operations. 'We had no more heart in the work after this diver's death', Mr V. Doeksen told us. Besides, only material of little value was brought up, though this was in great demand. The vogue for fashioning objects from *Lutine* material increased rather than decreased, and so on the occasion of his silver jubilee, Lloyds Committee presented King George V with an inkstand made from wood recovered from the wreck the previous year and sent to London. In the middle of September it was planned to attempt to raise the tower and set it down in a fresh position, it being explained that the season was drawing to a close. But Jonker & Son's product showed little inclination to follow its predecessor's example. It was not troubled by internal weakness and so when the *Texel* and the *Holland* tried to dredge the tower free in the first week of November, they were forced to leave it in its place – *saevis tranquillus in undis*. It was still standing there, proud and unimpaired, when the winter storms were spent and spring arrived on the scene. In April 1936 Tjebbe Stobbe, an engine fitter, took out patent rights in Rotterdam for his invention, a 'sand suction tank', which was to be constructed by the firm of P. Smit and in which a well-known inhabitant of Rotterdam was to have a financial interest – without issuing shares – and with which the then sixty-seven-year-old diver Jan van Drimmelen, shortly before dismissed by the Wijsmüller bureau in IJmuiden, was to do some diving. Van Drimmelen was full of courage. He told a reporter: 'A fortune-teller in IJmuiden once predicted that I would find the gold bars but that it would mean my death.' The prediction did not come true.

1936 was certainly a year for inventions. A German engineer and associate of Beckers turned up on Terschelling with an apparatus which can best be described as a large iron umbrella. But, so Mr Doeksen informed us, the object kept toppling over. For that matter, he was not able to experiment with it – which he did on board the *Texel* – until 26 June, since for weeks on end the wind had been blowing from the north, ruling out all activity above the *Lutine*. Just before the German started to work with the new apparatus a new diving tower was given a trial in Amsterdam (on 12 June). It was made to the design of W. van Wienen by the Amsterdam Droogdok Maatschappij (Amsterdam Dry Dock Co.) and called 'Van Wienen's sludge and sand-diving appliance'. The novelty of this invention was that the sand or sludge was not sucked away, so that fresh sand or sludge kept streaming in, but was pressed back. Van Wienen's tower was not tried out on the *Lutine*, although its backer, the contractor A. Volker of Scheveningen, did manage to get a concession out of Lloyds in 1937 on behalf of Van Wienen's Salvage Company, which ran until 13 April 1938. Impressive names, among them that of a Delft professor, were quoted as advisors in a communiqué dated 22 March, but on 12 December of that year Volker was obliged to inform the Ministry of Public Works that he was unable to make use of the permit granted him on 17 June 'for the dredging of sand in the Vlie Inlet and the placing of a tower and seamarks'. A quarter of a century later, Van Wienen was to come into the news once more in connection with the *Lutine*.

In the summer of 1936 it became clear that the game wasn't worth the candle. According to official information, Beckers suspended operations on 23 June, and in the second half of August the *Texel* and the *Volharding* removed the upper part of Beckers' second tower with a derrick and conveyed it to the harbour of Terschelling. It was intended that the lower part would then be lifted – Doeksen–Dros had taken this task upon themselves, too – and the two suction-dredgers were reinforced by the *Bornrif* for the purpose. The cone had already been slung from the derrick when on 14 September, on account of 'a heavy swell which suddenly developed' (as the contractors stated in the official correspondence), the tower fell from the derrick, damaged the latter severely, and landed, slings and all, in the water, this time not standing vertically but reclining comfortably on one side. It was a plain enough hint that this part of Beckers' invention wished for the time being to be left in peace, and was in fact preparing to hibernate. After abortive salvage attempts by Doeksen-Dros in the summer of 1937, during which there were only ten '*Lutin*e days', this rather hefty Sleeping Beauty was awoken somewhat hard-handedly by Prince Karimata in 1938. By then, the Beckers period was definitely past. The small, active and agreeable southerner was forgotten as well, and forgiven, it is hoped, for his optimism.

23 The *Karimata* in action

Anyone who had not made up his mind where he should spend his holidays in 1938 received an interesting tip via the advertisement columns of various Dutch daily newspapers. Bold type, in an advertisement of striking size, shrieked out at the wretch who didn't know what to do with his time:

KARIMATA

The biggest tin-dredger in the world!
A technical marvel!!

dredging for the *Lutine* gold
between Vlieland and Terschelling

Boat service to the *Lutine* site
if there are enough passengers

The advertisement was signed by the Doeksen Shipping Company of Terschelling and was an eloquent illustration of the art of buttering one's bread on both sides. For the shipping firm which sought to give seaside visitors a glimpse of the technical marvel was the very one that was earning a pretty penny thanks to the activities of the said marvel.

In the spring of 1938 the tin-dredger *Karimata*, built to the order of the Billiton Collective Mining Co. Ltd, reached completion in the shipbuilding yards of J. & K. Smit Ltd at Kinderdijk, which yards worked in co-operation with Heemaf Ltd, of Hengelo, and Sulzer Bros. of Winterthur. As the advertisement said, it certainly was the biggest dredger in the world: 246 feet long, 75 feet wide, 4,200 tons, fitted with three 600 h.p. diesel engines, capable of dredging to a depth of 98 feet and of handling 523 cubic yards of earth per hour with its 160 buckets, each of 88 gallons capacity. Deep-sea tugs were originally to tow the tin-dredger to the then Dutch East Indies, and with the weather conditions in the Indian Ocean in mind, the departure from the Netherlands had been scheduled for March or July to August. But if it departed in the spring, the tow ran the risk of bad weather on the journey to the Mediterranean. This fact, added to the circumstance that, as a result of the restrictions on the international trade in tin, the *Karimata* was not yet needed for tin mining and could well be made use of in her own country, probably persuaded the management to take the risk. Did their hopes run high? In an article, Van Capelle stated *after* the event: 'Thus up to the beginning of the year 1938 one was faced with the following situation: there were well-founded reasons for assuming that assets worth between eight and eighteen million guilders lay hidden at the *Lutine* site. There was no certainty on this score, so one had to take into account the fact that one might recover less, or even nothing at all. In other words: it was the same state of affairs as one is apt to encounter when one starts to explore an area with natural resources.'[43]

Before this exploration took place, various things had to be done: an investigation on the site of the wreck into the depth of water, currents, anchorage, the position of sandbanks and channels and the possibility for tugs to approach closely to the dredger and to take refuge at a safe mooring in case of bad weather; the drawing-up of a 'no cure, no pay' contract with Lloyds, the 'decretal salvors' and other rightful claimants to the cargo; the obtaining of the approval of the Ministry of Public Works and the Board for Crown Lands; the fitting-out of the dredger for this special job, the drawing-up of a contract with the firm of Doeksen, who were quite rightly considered to be thoroughly familiar with the conditions on the site and who would have to stand by with men and materials. All this was done before the *Karimata* left Kinderdijk for Terschelling on 4 June 1938.

The first buoy had been placed on 14 March, after which the *Texel* the *Volharding* and the *Neptunus* made a start with suction-dredging.

The tin-dredger had about a nine-foot draft, and the tug *Holland* (a predecessor of the present well-known sea-going tug) ten. The shallows which were a source of danger were found from the NW around to the east side. Winds from west through SE were therefore the winds which would drive the tin-dredger towards the shallows, should the hawsers break or the anchors cant. Now SE storms only occurred very rarely – so they calculated – and on the *Lutine* site they were land winds, which consequently did not raise a heavy sea. The same applied more or less to the south wind. But careful attention had to be paid to the SW and west winds, although it was fortunate that the SW wind blew partly along the cost of Vlieland and therefore did not cause such a heavy sea as a NW wind. The plan that the dredger worked with its head, i.e. with the bucket-ladder, facing SE, one advantage of which was that the dreaded north-wester would drive the dredger in the right direction.

In the meantime no half measures were tolerated. All the steel cables to which the *Karimata* was attached were 40mm thick and could take a strain of 100 tons, with the exception of the stern cable, which was originally 38mm thick and could take a strain of 86 tons. But after it had given way in the violent Force 11 storm of 27–29 June this cable was replaced by one of 50mm. The possibility of repeated 'flights' was also taken into account, for experts and 'people with experience', as they called themselves, had asserted that there would be no holding the *Karimata* even in Force 5 winds. Before long, however, experience showed that it was still possible to carry on normal working with a SW wind at Force 7.

The *Karimata* was given a bower cable mooring which lay 850 yards in front of the vessel and consisted of three stocked anchors one behind the other. The mooring of the stern cable consisted of a three-ton and a two-ton anchor with enlarged flukes and lay about 750 yards astern of the dredger. The port front cable mooring was about 437 yards away from the *Karimata* and consisted of a two-ton and a one and a half-ton anchor behind 80 feet of chain. The port rear cable was anchored in the same way. The starboard anchors weighed three tons, and the dredger possessed a gigantic emergency anchor weighing a couple of hundred tons in the form of its bucket-ladder with bucket-chain, thus ensuring 'calm amid the waves'.

Obviously, the Billiton company's involvement in such a spectacular fashion with the *Lutine* gold made a deep impression, even on sceptics and pessimists. *Lutine* stocks, which had slumped heavily after the contretemps with Beckers' towers, suddenly boomed, and even if expectations did not run very high, they were far more hopeful than on former occasions. Indeed, the very look of this salvage attempt spoke to the imagination: a mighty product of human ingenuity and industrial effort, brought into play not by impractical dreamers or financial speculators but by a reliable firm, was stationed over the carefully pinpointed site of the famous wreck, in full view of the islands and as brightly illuminated in the dark of night as if some fairy palace had risen from the depths of the sea. Well-known personalities showed an interest, a small army of newspaper men took up their quarters on Terschelling, and curious seaside visitors congregated in the little town of West Terschelling, down at the harbour and in the Nap Hotel.

On 9 June the *Karimata* was towed out to the *Lutine* site and at 6.30 a.m. the engines revved up and the buckets took their first bite at the sea bed, then fifteen to thirty feet below the surface. Small copper nails, pieces of copper and even fragments of a copper medal were brought up in the first days. The medal bore the image of a man's head, thought to be that of Frederick William III, King of Prussia. The legend on it read *Galle fecit*. Although the weather was not all one could have wished, the *Karimata* went on with her work, and on 12 June had the first silver coin, a Spanish piastre (*real*), to show on the floor of the riddling-drum. The coin, which was almost undamaged, bore the likeness of Charles IV on the obverse, round which the legend *Carolus IV Dei Gratia 1789* was clearly legible. On the reverse was the coat of arms of Spain, two lions and two castles surmounted by a crown and flanked by the two words of the device *Plus ultra*.

From time to time portions of Beckers' last tower, which was still lying right across the wreck, were brought to the surface coiled up – the dredger went through it as if it were a cardboard box instead of an obstacle made of manganese steel! It even brought up fragments of the three cubic metre concrete block to which the *Lyons* had been anchored in former days.

The *Karimata* carried on quietly with her work until on 27 June a violent storm came up out of the west, very soon reaching wind Force 11. It whipped the sea over the shallows into such a seething turmoil of waves that the *Karimata* had a very hard time of it. Too hard, for not only did the stern cable snap, but a bollard was torn out of the deck. 'The dredger was hanging on by the bow cable and could be towed to Terschelling only with great difficulty. It was what you could call a narrow escape,' said Mr V. Doeksen.

After repair the *Karimata* was towed back to the wreck on 14 July, after which it was possible for the second salvage attempt to begin. Among other things, the buckets, tirelessly dredging away, hauled up a great deal of ballast, various copper coins and also silver ones, a large copper kettle and an iron seal bearing the name E. I. T. Porter. They were seen as auspicious heralds of the coveted bars of gold. And indeed it did look as though trouble and expense, initiative and labour would be rewarded at this time.

It was at about two o'clock on the night of 28–29 June – the work being carried on day and night by three shifts of seventy-five men – when Messrs Doeksen and Van Capelle, asleep on board, were awakened by an excited voice through the ship's telephone, stammering that the first bar of gold had been found and that this time it was not just a joke or a rumour. Mr Doeksen, who naturally remembers what happened very clearly, told us: 'People did say afterwards that we had thrown that bar of gold into the sea ourselves, but I can assure you that we were elated at the find. I went ashore with it first thing in the morning, but Van Capelle said "I'm staying here. I'll be bringing the rest along soon." He was off his head. And I, too, thought the treasure would be coming up any minute. And so I put a board ready to stack the bars of gold on. When no more followed, there was naturally a great deal of disappointment, but I was still glad that we had found that one bar, for it showed we were over the right spot. I immediately took the dredger back fifty yards or so and started dredging six feet deeper and over a wider strip.' So went his account – a tale quietly related by a man who is not easily thrown off balance.

As far as the gold bar was concerned, it was found 60 feet away to starboard, measured 8 in. by 2½ in. by ½ in., weighed 7¾ pounds, was worth about 7,120 guilders and was immediately put into the hands of G. L. Bol, the Billiton Company's works manager on Terschelling, and who was accompanied by the Lloyds sub-agent H. Gongrijp. The find was taken to the Nap Hotel, where the flag was forthwith hung out. A great many ships in the harbour ran up flags too, and sirens made themselves heard. When the gold bar was cleaned, it was seen to bear the mark 2 F BB 57. It was soon established that no. 56 of this series had been found on 23 September 1858 and no. 58 on 14 November 1801, so it fitted in precisely between the two.

While Terschelling was still buzzing with rumours and excitement had risen to boiling point, the news was made known at Lloyds and caused no less of a sensation there. One journalist reported as follows:

'The report came in at about ten a.m. English time and was announced as good news. The clapper of the *Lutine* bell was struck

143

twice and the speaker made the announcement that a bar of gold had been found on Terschelling, weighing 7¾ pounds and worth about 7,000 guilders. Who could have dreamed that the same bell which left England with the *Lutine* on a cheerless day would let its metallic voice be heard once again 139 years later to tell of the good news about the *Lutine*? Everybody was happy and excited. People cheered and shouted hurrah!'

Gold was found again the following day, but this time it was no more than a large coin, dated 1797, while on the same day three cannon about eight feet long were recovered with their cannonballs, tied up with hempen rope that was still fairly sound. On 5 August another cannon followed, and yet another on the 6th. But the buckets did not bring a single bar of gold to the surface, though on and after 9 August they did bring up a so-called spade guinea (a coin of the period 1787–99 which took its name from the spade-shaped shield on the reverse), a French coin of 1786, a Spanish coin of 1793 and a Sicilian of 1735. Meanwhile discussions were going on about the reason for the absence of fresh discoveries of gold and silver bars. The conclusion reached – a fairly acceptable one – was that bar 2 F BB 57 had become separated from the great mass of bars during earlier salvage attempts and was buried under the sand, or else had simply fallen back into the sea while being hauled up.

This second salvage attempt lasted until 15 August. Then the *Karimata* had to put back into Terschelling Roads again to have some damage repaired, but a week later the buckets were groping over the *Lutine* site once more. There is little of importance to relate about this third period. Some further coins were fished up, including some gold ones, but no more pieces of wreckage came up to give an assurance that they were working over the wreck. With 1 October approaching – the date on which the contract with Lloyds expired – the Billiton Co. suspended operations on 12 September. The *Karimata*, which had by then worked over a territory 186 to 230 yards long, 77 to 142 yards wide and up to 66 feet deep, was towed back to Kinderdijk, and in July 1939 the gigantic tin-dredger left for its destination in the Dutch East Indies.

In the meantime the company was left to draw up its accounts. As appears from figures quoted by Bennendijk,[44] the following items were recovered in addition to the famous gold bar: 8 gold, 123 silver and 10 copper coins, 330 lbs of copper in the form of various kinds of nails, ship's sheathing and clamps, about 29,000 lbs of iron, about 660 lbs of lead, approximately 24 cubic yards of ship's timbers, about 1,000 cannon balls of varying diameters, a musket, a bayonet sheath, 13 copper uniform buttons with small anchors on them, a copper badge which probably formed part of some soldier's equipment and was stamped 'G R Per Mare e Terram', a small copper medal from London (1796) and a large and undamaged silver specimen with a Latin inscription, dated 1786, conferred by the University of Bonn. The cannon, five in number, have already been mentioned. Out of the total proceeds of these articles 12,038 guilders were due to the Billiton Company, against which stood the sum of 240,554 guilders for net costs. The costs of the entire undertaking amounted to no less than 442,554 guilders, of which the company recovered 189,995 guilders from insurance, as stated in its annual report for 1938, to which information it added drily: 'We need hardly add that we shall not be repeating the attempt.'

Van Capelle, the engineer, commented: 'Economically speaking, the undertaking was a failure. The gold present in the area worked over by the *Karimata* has been recovered, one need have no doubts on that score. It is a pity that there was so little.' Though opinions differ on this point (the view also exists that the gold, together with part of the wreck, has become widely scattered around the site) the engineer's conclusion that the undertaking was 'technically a complete success' will certainly not meet with contradiction. The company's annual report underlined this once more by stating:

> 'Technically the undertaking has come up to the highest expectations; it was possible to carry out the job completely according to plan. This is important in itself, since reclamation equipment involved here is far larger than any formerly supplied by Dutch shipyards. The exceptionally unfavourable conditions in which the work frequently had to be carried out provided ample opportunity for demonstrating its technical possibilities, and experience has been gained which will bear fruit in the future both for the tin industry and for Dutch dredger builders.'

Wasn't the price paid for these benefits rather on the high side, one will ask. Well, Terschelling had a tiptop summer as a result, the seaside visitors were kept entertained and sometimes thrilled, there was absorbing material for the journalists, and neither the Billiton nor the Doeksen Company went short of publicity. For the rest, the Dutch shareholders probably took their loss with the same impassive faces as the British underwriters in 1799 – both were able to stand a loss. Finally, the salvage attempt with the *Karimata* was the last raid on the tormentress's treasures. It was also the most spectacular and the most expensive one; and deserved a better outcome. The *Karimata* did nothing by halves. Should others ever wish to venture on a new salvage attempt – which is hardly likely – it is possible that they may find gold, but they will certainly not come across any remnant of the *Lutine* herself. The vessel foundered in 1799, but in 1938 she was destroyed.

24 A floating saucer

It is now more than a quarter of a century since the *Karimata* worked over the wreck without the desired result, and in all these years no further salvage attempt has been made, although in 1956 the papers did speak of new plans. How could it be otherwise? The tantalizing question still remains: if the *Lutine* was so richly laden, what has become of all those millions? These very facts – that the accounts do not tally, that considerable sums in coin are missing, that in the numbered series of gold and silver bars a great many numbers are still missing – are the reasons why despite the very small measure of success during the last eighty years of 'treasure fishing', people have from time to time fallen so much under the spell of the mysterious bullion ship that they wanted to have another go at finding the rest of the treasure.

And so in June of 1965 the press published the photograph of someone already known to us, the sixty-year-old engineer W. van Wienen, of Bussum, managing director of the Netherlands Diving and Salvage Co. Ltd of Flushing, founded in 1953. Van Wienen had worked out new plans for the umpteenth *Lutine* enterprise. The difference now was that this naval architect declared that it was not his wish to investigate the wreck so much as the route the *Lutine* had followed shortly before she sank. He reasoned that the possibility existed that some hundreds of yards away from the place where she finally came to lie, the vessel struck a sandbank with such force that her side was holed at the level of the bullion room, through which a large part of the gold and silver disappeared into the sea.

This would provide an explanation of why the *Karimata* had thoroughly investigated the area of the wreck without noteworthy success. It would also tie up with the somewhat remarkable position of the wreck shown by Taurel's research, namely, with the stem facing the direction from which the storm blew on that fatal night. Had they then still managed to manoeuvre the battered vessel? We shall certainly never find out, for the Bussum engineer's plans remained mere paper plans and his saucer – not a flying but a floating one – never got beyond the model stage. The inventor had designed a hanging sludge and sand-diving apparatus, which was attached to a derrick mounted on a circular float. This float, which was 115 feet in diameter, would have the advantage, the inventor informed the press, that the force of the tidal current would find no point of impact, so that it would not be necessary to keep shifting the salvage ship. 'The diving apparatus, which weighs 40 tons – a hollow tube with a workroom at the bottom in which four persons can work – is kept under an effective pressure of 1.6 atmospheres, which prevents seawater from flooding in. It can go down to a depth of 80 feet. The divers enter the diving tube in the sluice chamber and can lower themselves down with a lift.' A special feature was said to be that on their floating saucer the twenty-six men would have little contact with Terschelling.

Van Wienen estimated the value of the bullion still present as 42 million Dutch guilders – in which estimate creeping inflation has played its part – and he calculated the costs of the salvage work at 1.5 millions. Taking 100,000 guilders into account for unforeseen expenses, an issue of 1.6 millions was opened on 27 October 1956. Unfortunately, only a month later the inventor was forced to come to the melancholy conclusion that the man who was to act as his financial adviser was a swindler who had been 'obsessed for years by the gold fever'. This man was stupid enough to appear at a press conference, where he was promptly unmasked. His case – a different matter incidentally, from that of the new *Lutine* enterprise – ended in March 1958 before the Amsterdam court with an unpleasant sentence for the swindler. In August of the preceding year the manager of the Netherlands Diving and Salvage Co. Ltd had been declared bankrupt, thus putting an end to the plans for trying to solve the riddle of the sunken treasure.

Anyone reflecting on the outcome of this affair cannot avoid the impression that here the tease was playing her last card. But will it be the final one? We have had sight of correspondence with Dutch government departments since the Second World War dealing with no less than five fresh projects, some including inventions, with reference to the *Lutine,* the first of which dates from 1949 but the last from as recently as 1961! In 1949 a man from Augsburg cherished salvage plans: in 1953, in return for reasonable compensation, a man from Antwerp was prepared to place his project at the disposal of the Dutch state, 'which would considerably alleviate the distress of the Dutch people' (this was an allusion to the great flood disaster in February of that year); in 1956 a sea captain from Rijeka in Jugoslavia showed an interest; in 1958 a contractor from Duisburg-Hamborn tried to get in touch with the Dutch government and in 1961 an inhabitant of Schweinfurt in West Germany followed his example. So the *Lutine* is very far from being forgotten. Besides, will there ever be a definite farewell? Lloyds of London keep the tradition – cult, if you like – alive, and who knows how often a boy from the mainland visiting Terschelling will ask what cannon are those that raise their barrels to the sky down by the harbour? Time and time again there will be a father who will take his son off to a high sand dune or to the gallery of the lighthouse and point out to sea. 'There,' he will say, 'that's where it was. There where the white surf is breaking in the distance, that's where the *Lutine* went down. A ship laden with pure gold!' It sounds like a fairy tale, and that is as it should be. As in every true saga of the sea, imagination can no longer be divorced from truth – a truth so deeply buried under sea and sand that it speaks all the more powerfully to the imagination. A truth which even manages to elude facts and figures, records and reports, and has never come wholly to light.

25 Lloyds and the *Lutine*

In the very earliest reports about the ship there is already mention of Lloyds, whose name was permanently linked with the wreck's fortunes once it was assigned a half share of the value of its contents. This, as we saw, was in 1823, nearly a quarter of a century after the insurers had had to pay out, the *Lutine* having been written off as a total loss. Since then Lloyds has become the guardian of the *Lutine* tradition and even in its most recent publications it does not fail to draw attention to its possession of the bell, table and chair, all of which play a part in the *Lutine* ceremonial. It is also worth mentioning that Lloyds white yacht, which habitually pays a visit to the Netherlands during its voyages, bears the name *Lutine* – as a matter of course, we would almost venture to say. There is therefore every reason to retrace the history of this financial organization whose name stands for reliability, so much so that all manner of businesses outside Great Britain connected with shipping and insurance have adopted it themselves. Think, for example, of Rotterdam Lloyd in the Netherlands. Lloyds is an insurance market where, with few exceptions, any insurable risk can be placed with Lloyds Underwriters through their brokers; it is also a society incorporated by Act of Parliament in 1871, and the world centre of marine and other information.

A long period of evolution was necessary to get this far. In fact, we find no brilliant financier at the beginning of Lloyds' story, but a respectable coffee house proprietor, though in all justice it must be said that he would probably be called a 'slick customer' in our day. This man was Edward Lloyd and his coffee house stood in Tower Street, in the heart of the City of London. According to an advertisement, Lloyd already owned this coffee house in 1668. In 1692 he moved to Lombard Street, on the corner of Abchurch Lane, in the middle of the business quarter of those days. And, as in Tower Street, brokers met one another here, too, to talk business over a cup of coffee or something stronger. Edward could do more than make coffee and pour out gin. He managed to make his house a centre for all who had an interest in shipping and shipping insurance. He must have been a man of foresight, for in September 1696 he began publishing a news sheet entitled *Lloyd's News*, to supply information about shipping, a matter which was as important as it was difficult in that age, with its lack of good communications. After being banned as a result of a dispute with the House of Lords, the news sheet reappeared in 1734, this time as *Lloyd's List and Shipping Gazette*. It still exists as such and is London's oldest newspaper.

The paper should not be confused with *Lloyd's Register*. This goes back to the registers or lists of ships, written with a quill pen, which passed from hand to hand in Lloyd's coffee house and which later appeared in print. Ships were entered and classified according to type and size. It is known that there was a printed register in 1760. After a reorganization in 1834, this grew into the great organization known as Lloyd's Register of Shipping, at present housed in Fenchurch Street. It is a separate company, on whose board the Corporation of Lloyds is well represented – the latter's chairman, for instance, always having a seat on it. The body which publishes the Register is popularly known as the Society.

Nowadays Lloyds is a company whose members are known as underwriting members of Lloyds and transact insurance business on their own account and at their own risk. It was originally what one might call a club of underwriters, who had their own accommodation – from 1774 onwards in the former Royal Exchange – but could not act for themselves, for example in legal proceedings. Even the receipt of the money from the *Lutine* from the salvaging in the years 1847–61 was attended by difficulties. Whose money was it, after all? Parliament did not decide to recognize Lloyds as a corporate body until 1871.

In order to acquire this corporate nature the members had to send in a list of the business they had on hand. The list consisted of twelve sections, four of which had reference to the *Lutine*. So Lloyds must have been very closely concerned with the *Lutine*, although there are no exact details of the size of the sums underwritten by its members. It has often been said that the relevant documents were burned in the fire in the London Stock Exchange on the evening of 10 January 1838, but Lloyds' information bureau wrote us that it did not know how far this was true, that 'very much information' was lost in the fire.

When in 1844 the Exchange rose from its ashes, Lloyds was housed in the new building until the company moved into a building of its own in Leadenhall Street in 1928, which was vacated in 1957 when

Lloyds took up its residence in a large new building in Lime Street.

Anyone who should wish to embark on salvage attempts on the *Lutine* is obliged to deal with Lloyds. For Lloyds have possessed half the rights to the wreck since 1823, and so have always demanded a percentage of the proceeds of each successive attempt to recover the treasure, although this percentage has varied. In exchange, Lloyds have granted permission for salvage attempts to be made. Evidently the wreck of the *Lutine* has not yet been written off for good. Even as recently as 1963 the Dutch agent of Lloyds was obliged to inform us that the company was not prepared to grant us unlimited access to the records concerning the *Lutine* which are at present in Amsterdam. One of the reasons given was the confidential character of the greater part of the material as well as the fact that 'in the nature of the case, hope has not yet been abandoned that gold may still be recovered.' Thus, in Lloyds' eyes the *Lutine* has not yet become a thing of the past, and there may still be a sequel to her story. We hope it will not be taken amiss of us that we did not wish to wait that long before writing this book, to which, after all, a supplement can always be added . . .

26 Relics of the *Lutine*

Souvenir hunting – the collecting of tangible keepsakes connected with famous persons, affairs or places – has been with us for a long time, and as soon as there is talk of attempting to recover parts of the *Lutine*'s cargo we hear of the raising of objects whose monetary value was not great, but which people nevertheless prized as souvenirs and sometimes even venerated as relics. When, on 14 October 1814, Pierre Eschauzier, after twelve attempts to 'find out whether the place where the money lies was uncovered' hauled up a gold Spanish piastre overgrown with weeds and shells, proof that his goal had been reached, he wrote in his report: 'N.B. this sent to the President of the Councils and Auditors-General of the Crown Lands to be placed in His Royal Highness's Cabinet as a curiosity.' Apparently William I had built up a collection of curios, as had become fashionable in the Netherlands – especially when trade with the East Indies developed in the seventeenth century. Such a collection was called a cabinet or museum of curiosities. A good many further *Lutine* souvenirs followed this gold coin – coins, nails, cannonballs, pieces of wood – and there is no point in trying to trace the present whereabouts of all of them. Some have ended up in museums or with private individuals and organizations that were engaged in the salvage operations. Lloyds of London of course have a thing or two.

The most famous nowadays is the ship's bell, 'to many people symbolizing Lloyds', as a Lloyds publication puts it. It was on 17 July 1858 that this bell, bearing the date 1779 (the year *La Lutine* was built), the coat of arms of the French monarch and above it the words 'Saint Jean', was hauled up. The object, weighing 106 lbs and $17\frac{1}{2}$ in. in diameter, was taken to the rebuilt Royal Exchange in London and hung from an ornamental wrought-iron frame.

It was removed with this decorative frame to Lloyds' new building, opened in 1957, where it was placed above the rostrum, from which the mechanical caller calls out the names of brokers whom colleagues wish to contact. This rostrum stands in the centre of the south gallery of the underwriting room. When Queen Elizabeth II laid the first stone of the new building in November 1952, and when the Queen Mother opened it on 14 November 1957, the *Lutine* bell was solemnly rung twice. Otherwise its bronze is only sounded to obtain silence in the underwriting room whenever there is important news to announce, once for bad news and twice for good. In the past this news usually concerned ships that were overdue, but in our day it seldom happens that ships come in late or are missing, although the bell was rung twice on 8 July 1958 and on 29 October 1959, to announce the safe arrival of ships which were overdue. And after Sir Winston Churchill died on 24 January 1965 and Lloyds took part in the general mourning, the *Lutine* bell was rung at noon on the following day in order to call for a few moments of silence. Churchill was an honorary member of Lloyds and his wife is the daughter of the late Col. Sir Henry Hozier, Secretary of Lloyds from 1874 to 1906. According to the photograph in *Lloyds Log* of February 1965, at this commemoration in the underwriting room a large congregation paid a standing tribute to the great statesman's memory.

The name of Churchill leads one's thoughts automatically to the Second World War. We learn that in those days the *Lutine* bell was even mixed up with politics. An amusing item in *Lloyds Log* for March 1965 tells how the Anglo-German radio propagandist known to the English as Lord Haw-Haw asserted in one of his speeches before the Nazi microphone that during the war the bell rang non-stop at Lloyds to announce the vast number of losses at sea! In fact, the *Lutine* bell rang only once in those years and that was when the German battleship *Bismarck* was sunk. It rang only once, for although it was good news for the British, tradition insists that the bell should be rung once, not twice, for the loss of a ship at sea and the rule was adhered to then as well.

Another part of the bullion ship which also does service in Lloyds' building speaks less to the imagination. This is the rudder, which was raised in 1858 and on whose metal bands the name *La Lutine* was incised. This item was also taken to London, where a table and chair, ornamented with carving in the fussy style of those days, were made from it. These are now in Lloyds' writing room and are used by the chairman when he addresses the annual general meeting of members. Finally, various items are to be found in the library, most of them objects recovered in the years 1858–60 and varying from a hammer to a rusty watch, from guns to cannonballs, coins and badges and similar trifles.

Cannon have always been very much in demand as well. Lloyds received the first two in 1886 and 1888 respectively. One of them was offered to the City of London, to be placed in the Guildhall, the other was accepted by the Queen and later sent to Windsor Castle. A third cannon from the *Lutine* is at present in the grounds of Lloyds' sports club in Essex. Dutchmen, however, need not travel so far to see a couple of *Lutine* cannon, for in Holland they soon formed objects of interest in various places. At the beginning of this century two could be seen in the little Frisian port of Makkum, but after the closing down of the owner's firm of Maas, these cannon became the property of a certain Klaas Tjebbes of Workum and were afterwards removed to a country seat on the River Vecht.

According to information received from Mr N. S. Binnendijk, anyone in the Netherlands wishing to see cannon from the *Lutine* can do this in two places in Amsterdam: in a cellar belonging to the Stedelijk (Municipal) Museum and in the garden of Mr Alfred Schröder's house along Herengracht. More can be seen on Terschelling, where two specimens stand on an eminence close to the harbour of West Terschelling, where they flank a ship's anchor which does not come from the *Lutine*; another is stationed in front of the museum Het Behouden Huys, and yet another was placed in front of the Nautical College in 1963. The cannon down by the harbour on Terschelling were recovered by the *Karimata* in 1938.

Just as these cannon perpetuate the *Lutine* tradition on Terschelling, so people entering the Lloyds' building in London will always be introduced to the disaster and the riddles surrounding the *Lutine* by virtue of the bell, table and chair. The inscriptions on the furniture tell everything one would wish to know. On the table, the legs of which are carved in the shape of dolphins, one reads:

'H.B.M. ship *La Lutine*
32 gun frigate
commanded by Captain Lancelot Skynner, R.N.
sailed from Yarmouth Roads
on the morning of the 9th October 1799
with a large amount of specie on board
and was wrecked off the island of Vlieland
the same night when all hands were lost
except one man.

The rudder of which this table is made and the rudder chain and bell which the table supports were recovered from the wreck of the ill-fated vessel in the year 1859 together with a part of the specie which is now in custody of "the Committee for Managing the affairs of Lloyds". Decr. 1866.'

This inscription refers to an earlier site of the ship's bell, for in 1876 it stood partly entangled in the massive rudder chain, on the footrest of the handsomely carved table. The inscription on the armchair – high backed, impressive in appearance and richly carved – reads:

'This chair
is made from the rudder of
H.B.M. frigate *La Lutine*
which sailed from Yarmouth Roads
on the morning of the 9th Octr. 1799
with a large amount of specie on board
and was wrecked the same night off
the island of Vlieland, when all on board perished
with the exception of one man.

The rudder was recovered from the wreck 1859, having been submerged for 60 yrs.'

Here, too, 1859 is mistakenly given as the year in which these mementoes were hauled up. The facts are that the ship's bell was hauled up on 17 July 1858, and that the rudder was raised on 18 September of that year.[45]

27 Numismatics and the treasure

There is every reason to devote a short chapter to the *Lutine* treasures as seen through the eyes of a numismatist. This task has been made very much easier for us by the British numismatist J. D. A. Thompson, of the Ashmolean Museum, Oxford, who wrote about the *Lutine* coins in the *Numismatic Circular* of October 1963, with special reference to those recovered in the years 1886–1938. According to this author, the circumscriptions dating from the period before 1886 are so unclear that numismatics can do little with them. In fact, we usually only hear of Spanish piastres, and there is now and again mention of a pillar piastre and also of a Louis d'or. The expert may find such designations too general, but the layman will be content when he knows that piastres, so Dr H. Enno van Gelder, director of the Royal Numismatic Collection, was kind enough to inform us, are eighteenth-century Spanish coins the size of a florin and officially known as *reales de à ocho*, or reales. Most of them would have borne the names of Charles III and Charles IV. The name pillar piastre (also pillar or pillar dollar) indicates a type of coin common in South America which has the pillars of Hercules on the reverse. Dr Enno van Gelder had more difficulty with the 'golden Spanish piastre', which we also came across on one occasion. It is possible that this very unusual combination indicates the Spanish gold eight escudos coin officially known as the onza. It is the same size as the silver Spanish piastre but, of course, far more valuable. A Louis d'or is a gold coin bearing the head of Louis XV or XVI of France.

Lloyds granted Thompson access to the list of finds. On going through them, his impression was that apart from the Spanish 'dollars' (piastres) very few of the coins identified by him formed part of the official *Lutine* treasure. The British, French and Italian coins probably belonged to the crew and passengers. Most of the gold coins were valid commercial currency in the Free City of Hamburg, while the copper halfpennies bearing the figure of George III may similarly have been used for small change. Thompson arrived at the following categories and numbers for the period he studied.

Spain and Spanish America

golden piastres (pistoles, onzas, doubloons)	8
half piastres (escudos)	1
quarter piastres (half escudos)	6
eighth piastres (escudillos)	6
silver piastres (daalders)	2205
half piastres (4 reales)	20
quarter piastres (2 reales)	17
eighth piastres (1 real)	2

Although the coins from Spanish America are not listed separately, it is highly probable that both Mexico and Peru are represented. Charles III (1759–88) and Charles IV (1788–1802) are both represented.

Sicily and Naples

The denomination of these coins is difficult to determine, but the small size is probably the oncia of Ferdinand IV (1759–99), and the larger variety might be the doppias of Charles III of Bourbon (1735–59); a coin bearing his likeness was certainly found in 1934, according to Thompson.

Sicilian gold coins (oncias?)	9
larger gold coins (doppias?)	4

France

double Louis d'or	38
single Louis d'or	51

Great Britain

guineas, including 'spades', dated 1791 and 1793 and one of an undetermined type dated 1790	15
half guineas	6
an unknown number of copper George III halfpennies	–

Germany

silver primatus daalder (not determined; perhaps a daalder, or half crown, of one of the Archbishops of Salzburg)	1

Unspecified coins

large gold coin dated 1797	1
silver	11
copper (mostly English?)	24

$$\overline{2,429}$$

One more species of coin can be added to this survey, namely a Russian Catherine II copper five kopec piece, dated 1793. This at all events is how Dr Enno van Gelder identified a coin which appears on a photograph in our possession showing coins recovered in the thirties of this century. The coin displays a florid monogram with the date 1793. On the same photograph the copper George III halfpennies referred to above are apparently shown, too, with the legend GEORGIUS III DEI GRATIA.

When we drew the attention of Mr Thompson to this find, the numismatist wrote that the presence of such a Russian coin did not surprise him. He surmised that the coin would have belonged to someone on board. The Russian army was billeted in Den Helder, so that there is no reason for arguing that a Russian coin could not have found its way on board the *Lutine*. 'And this is an argument for supposing that Skynner broke his journey for a couple of hours before going on to Cuxhaven.' This argument, in our opinion, rests on too flimsy a foundation, since no evidence at all can be found that the *Lutine* did in fact put in at Den Helder or Texel between Yarmouth Roads and the IJzergat. But we have already discussed this point.

Since the original Dutch edition of this book appeared, we have ascertained that coins raised from the *Lutine* were put on public sale in 1858. In the Provincial Library of Friesland at Leeuwarden we came across a catalogue referring to the sale of ancient and contemporary medallions and coins 'which includes a set of gold and silver coins lately recovered from the ship the *Lutine*, lost in 1799.' The sale took place on 18 February 1858 in the Huis met de Hoofden in Amsterdam, we read. The items in the catalogue were nos. 721 to 762 and included 29 Spanish and 12 French coins, one United States dollar and also a 'Factory Coin from Norwich'. The Spanish coins were mainly piastres and the French Louis d'or.

Acknowledgements and bibliographical note

In writing this book grateful use was made of the *Lutine* archives belonging to the Eschauzier Family Society; of the relevant documents present in the town hall at Terschelling; of file no. 262,351 (Wrak *Lutine*) in the offices of the West Friesland Department of the Rijkswaterstaat (State Department of Dykes, Roads, Bridges) at Leeuwarden; of reports in national, regional and local newspapers; and of a large number of works in Dutch, English, French and German, all of which are listed on pages 193–6 of the Dutch edition. Only the works in English are mentioned below. Any reader wishing to consult the other titles is referred to the Dutch edition, particular attention being drawn to the works of W. H. ter Meulen.

Anonymous, A Sketch of the History of Lloyds, reprinted from *Lloyds Calendar*, no place or date of publication.

Anonymous, Story of a Nail, *100 A1* (a bulletin published by *Lloyd's Register of Shipping*), no. 10, 1962, 24–26.

N. S. Binnendijk, Story of the *Lutine* salvage, *The Log*, 1934 and 1938.

J. J. Fletcher, *Some Materials for The True Story of the Lutine*. Copy of typed version of the text in the keeping of the Frisian Marine Museum at Sneek, Friesland.

H. Hozier, *H.M.S. "Lutine"*, printed report for Lloyds, not on public sale, 1895.

Frederick Martin, *History of Lloyds and Marine Insurance*, London 1876.

K.B.M., The Bell at Lloyds, *Lloyds Log*, March 1965.

C. Nepean Longridge, *The Anatomy of Nelson's Ships*, London 1955.

J. D. A. Thompson, Notes on the *Lutine* Treasure, in *Numismatic Circular* no. 10, October 1963.

J. Steven Watson, *The Reign of George III (1760–1813)*, Oxford 1960.

Arthur Young, *Nautical Dictionary*, London 1863.

Notes

1. Gerben Colmjom, *Hoe het oude Terschelling zich vernieuwde* (How old Terschelling was reborn), 1950, pp. 107–8.
2. G. Knop, *Brandarisflitsen* (Brandaris flashes), p. 5.
3. J. A. Acket, *Het eiland Terschelling in zijne eigenaardigheden geschetst* (Curiosities of the island of Terschelling), 1882.
4. ibid. p. 45.
5. Several sources: *Gentleman's Magazine*, November 1799; James' *Naval History*; a calculation by Lloyd's agent John Mavor Still, dated 1858; *Leeuwarder Courant*, 13 November 1799, which got its news from London via Paris.
6. *Algemeen Handelsblad*, 27 April 1857, in a report of a meeting of the Royal Netherlands Institute of Engineers.
7. The author had written these lines, based on correspondence mentioned in Fletcher's *Some Materials*, before he was able to consult the chapter on the *Lutine* in Martin's *History of Lloyds* in which a similar line of argument is in part pursued.
8. Brand Eschauzier, *Nadere mededeeling aangaande den arbeid op het wrak der "Lutine"* (Further information regarding work on the wreck of the *Lutine*), 1861, p. 17.
9. German version only consulted: *Die Bergung des Gold- under Silberschatzes aus der "Lutine"*, 1928.
10. Vincent Nolte, *Vijftig jaren in de beide halfronden* (Fifty years in the two hemispheres), Middelburg, 1855, p. 21.
11. Erwin Wiskeman, *Hamburg und die Welthandelsgpolitik von den Angfangen bis zur Gegenwart* (Hamburg and world trade policy from its beginning to the present day), Hamburg, 1929.
12. H. Reinke, *Hamburg*, Bremen, 1925, a brief history.
13. Also in *Gentleman's Magazine*, November 1799, p. 896.
14. From an article commemorating the *Lutine* in *Lloyd's List*, 25 November 1949.
15. From a letter from E. B. Merriman to Lloyds and noted by Hozier in his booklet *H.M.S. "Lutine"*, 1895, p. 2.
16. Lieutenant James Anthony Gardner (1770–1846), in *Blonde*, 32 guns, from 1798–1801. From *Recollections*, written 1836, published 1906.
17. *Lloyds Evening Post*, 21, 30 October 1799; *Hamburgischer Correspon-*

dent, 12 November 1799; *St James Chronicle*, 24–26 October 1799; *Bell's Weekly Messenger*, 27 October 1799.
18 J. J. Fletcher (while researching in 1895), information from the Weinholt family; also W. G. Stitt Dibden, *Postal History Society Bulletin*, May–June 1965.
19 Captain Portlock is referred to in the vice-admiral's letter (Chapter 6) as commanding the sloop *Arrow*, 28 guns. In company with *Wolverine*, 12 guns, he was keeping watch to the east of Vlie. See Hollema, *Geschiedenis van Nederland ter zee* (Naval history of the Netherlands), III, p. 423
20 Steel's *Naval Chronologist*, 1806, p. xxv.
21 B. Eschauzier, *Further Information*, 1861; also J. D. A. Thompson, Notes on the Lutine Treasure, *Numismatic Circular*, LXXI, 10 October 1963.
22 Van Veen, *Onderzoekingen in de Hoofden* (Research into the Moles), thesis, 1936.
23 Christened in Enkhuizen 1760, held office on Terschelling until 1808, died at Vollenhove on 12 April 1817. Son of Pieter Frederik Robbé, died 1797.
24 As appears from a letter from the Dutch Ministry of Home Affairs dated 2 December 1857, no. 103.
25 (see 24 above) The Minister's letter of 1857 contradicts this, saying, 'immediately after the wreck the Dutch Navy took possession of the stranded frigate, confiscating the anchors, cordage, cannon and mountings.'
26 Born Rotterdam 1772, son of Jean Pierre Eschauzier, died 1837. Chief Receiver of Wrecks until 1814, sheriff until 1831, burgomaster of Terschelling until 1837.
27 Dr Hamel, *Beschrijving van de duikerslok en eener nederdaling met dezelve* (Description of a diving bell and of a descent with the same), on Rennie's bell.
28 Printed in full in Martin's *History of Lloyds*, p. 199.
29 *Leeuwarder Courant*, 12 May 1843, report of activities on 2 May.
30 See also *Algemeen Handelsblad*, 1 August 1933, an interview with the diver Drimmelen who had worked on the shell-dredgers about 1885 and had known Wijker.
31 Brand Eschauzier, *Communication concerning the construction and fitting-out of the "Hollandsche Duiker"*, 1859.
32 Hozier (1895) credits Englishmen with recoveries in the period 1857–59, as does J. D. A. Thompson (1963).
33 F. P. ter Meulen, *Willem Hendrik ter Meulen* (11 January 1830–17 June 1901) *in zijn werkzaamheid voor de "Lutine" geschetst* (A Portrait of W. H. ter M. and his work on the *Lutine*), 1907.

34 *De Ingenieur*, 1901.
35 *Het bergen der schatten uit de "Lutine"* (The salving of the treasure on board the *Lutine*), 1896, booklet.
36 Prince Hendrik himself thought up a device to remove the sand over the wreck: a pier made of piling. It was considered too dangerous and never tried out.
37 Nieman was a notably skilled and courageous diver and seaman. In 1889 he had had his fishing boat towed out of Maasluis by a tug to go to the aid of the British vessel *Yoxford*, and succeeded in rescuing her entire crew of 22 men, after the lifeboat crew had refused to go out. See C. J. M. van der Hidde, *Bergers* (Salvors), Leiden, 1946.
38 Fletcher's letters were mostly anecdotal, though he did collect material mostly from contemporary journals and records, for a book, published in the form of *Some materials for the true story of the "Lutine"*.
39 Story of the *Lutine* salvage, *The Log*, August 1938.
40 Elderly inhabitants of Terschelling say that experiments with electric light were made at the wreck in 1903–4.
41 *The Log*, August 1938.
42 Typewritten text of an address, in *Lutine* archives.
43 *De Ingenieur*, no. 54, 20 January 1939.
44 *The Log*, October 1939.
45 Brand Eschauzier, *Further Information*, Appendix I, 1961.

Index of Names

Acket, J. A. 14, 61
Altena, P. 98, 102
Arntzenius, J. O. H. 99
Asmus, J. P. 50
Aufrere, Charles Gastine 22, 38

Bakker, G. 83
Beckers, Frans 126–138
Beer, Jan de 127, 136
Bell Bros. 69, 73, 74
Bell, William 71
Bethel, John 74, 75
Billiton Collective Mining Co Ltd 139–145
Binnendijk, N. S. 73, 112, 116, 125, 144, 155
Blo(e)m, Hendrik 59
Blom, Willem 41, 42
Blommendal, A. R. 99
Bol, G. L. 143
Bourdrez, J. J. L. 113
Brand Eschauzier, Jean Pierre 27, 28, 39, 81–84, 88–92
Brandenburg, Johannes 59
Bridges 26
Broc, Maurice 135
Broecke, D. S. van den 108
Brune 44
Buren Schele, Van 105, 115
Buijskes, A. A. 50
Burgt, J. H. van der 49, 51, 53, 54
Bijl 83
Bijlsma 126, 136

Capelle, A. E. F. van 140, 143, 145
Catherine II 159
Charles III 157, 158

Charles IV 142, 157, 158
Charles III (Bourbon) 158
Chatillon, Duc de 38
Churchill, Winston 154
Cochrane 67
Colmjon, Gerben 13, 14
Conyngham, F. 71
Cook 25
Cop, H. 108
Cruso, H. 67, 69

Dawson Warren, R. 36
Deinse, J. H. L. van 108
Denny Sargent, Th. 78
Diesen, G. van 95, 99
Doeksen, G. 30, 121, 123–138, 143, 145ff
Doeksen, Jan 74
Doeksen, Volkert 142
Downs, H. 79
Dros, Albert 30, 103, 116, 120, 123–138ff
Dros, P. 30, 102, 120, 123
Drijver 90, 91
Drimmelen, Jan van 130, 137
Dumouriez 20
Duncan, Henry 25, 34, 35, 44
Duras, Victor Hugo 126

Eijk, J. A. van 99
Eliason, d' 34
Elizabeth II 154, 155
Ellis, Peter 69
Ellis, T. 106
Eschauzier, Otto 83
Eschauzier, O. J. A. 74
Eschauzier, Pierre (jr) 83

165

Eschauzier, Pierre (sr)   63-76, 84, 90, 92, 118, 125
Eschauzier, Wilhelm Samuel Louis 81, 98
Eschauzier-Uitenhage de Mist, Mrs 90

Ferdinand IV   158
Freyer, H. J. de   40
Figée   107
Fletcher, Johan J.   38, 108-110, 112, 114
Frederick, Duke of York   25, 44
Frederick William III   142

Gardiner, Charles A. P.   116
Gardner, James Anthony   36, 44, 45
Gelder, Van   83
Gelder, H. Enno van   93, 157, 159
George III   158, 159
George V   136
Geuns, J. van   78, 79
Goldberg, J.   44
Goldschmidt, H. S.   134
Goldsmid, B.   39
Goldsmid, N.   39
Gongrijp, H.   143
Gorter, J.   99
Goudriaan, A. F.   51
Grinwis, P. J.   51

Hakvoort, Gerrit   81
Hardcastle, I.   104
Hasselt, C. van   100
Haw-Haw, Lord   154
Hecking Colenbrander, P. A. van   119-122
Heemaf   139
Heinecke   28, 89, 90
Hendrik, Prince   52, 99, 100
Hill, E.   79
Hofmeester   74
Hood, Samuel   21
Hoog, A. van der   88
Hoult, W.   69
Hozier, H. M.   27, 29, 30, 154
Huet, A.   113

Istemaas, August   130-131

Jaffrey   110
Jongebroer, C.   65
Jong van Beek en Donk, J. O. de   120
Jonker, H.   133, 136

Kammen, Johannes Oene van   68
Keizer   59
Kerkhoven, Jacob Pieters   67, 69
Keulen, J. van   60
Keulen, Sjoerd van   81
Kikkert   83
Kinipple, W. R.   29, 108, 110, 112, 114, 115
Kloes, J. A. van der   129, 132
Knop, G.   13, 14
Kooiman, Trijntje   127
Kraijenhoff, C. R. T.   44, 99
Krul, P. T.   90-92, 96, 99
Krupp   132

Lloyd, Edward   150
Louis XV, XVI   157

Maas, Willem   102, 103, 111
Martin, Frederik   30, 35, 55, 63, 73, 76
Matthes   99
Merriman, E. B.   27
Mettinga, E.   136
Meulen, Arend   100
Meulen, F. P. ter   93, 101, 109
Meulen, W. H. ter   52-53, 85, 90, 93-113, 119, 160
Monckton, J.   22
Montgomery, Walter   38
Montmorenci, Duc de   38

Nap   142
Napoleon Bonaparte   21
Nepean   35
Neugebauer, Maximilian   122
Nieman, L.   103, 105
Nieuwenhuisen, J. F.   98
Nijkerk   83
Nolte, Vincent   32

Norton   74, 75
Nijgh, W.   120

Ortt, H. J.   66
Oudraat, J. C.   84

Panteydt, Teunis   82
Parish, John   34
Peel, Robert   117
Pijnappel, Menso Johannes   107
Porter, E. I. T.   143
Portlock   37, 41, 43, 55
Pottenga, S.   74

Reedeker, Meester   13
Reedeker, D.   98, 110
Reedeker, Jacobus   65, 68, 69, 77, 85, 88, 89
Reedeker, S.   81
Rennie   66, 67, 75
Reus, J. T.   86, 87
Reynders, J. A. H.   40
Rijn, K. van   98, 101
Robbé, A. A. de Vries   59
Robbé, F. P. (de Vries)   28, 50, 56, 58–61, 63, 80
Roest, Van Galen en   88
Roever, J. C. de   116
Rotgans, Gerrit Siebes   38, 40, 74, 81
Rundell   26
Ruyter, Michiel Adriaansz de   43

Scott, John   114, 115
Schröder, Alfred   117, 124, 125, 155
Schutte, Gebr   88
Serrurier, L. J. J.   68, 75, 78
Skynner, Lancelott   22, 26, 35, 36, 38, 41, 45–47, 155, 159
Skynner, Lawrence   22
Smit, J. and K.   139
Smit, P.   137
Sperling, Martijn   130, 131
Sperling, Matthijs   106
Sperling, Wouter   106
Stang, Theodoor   107
Stieltjes   95

Still, John Mavor   24, 91, 92
Stobbe, Hendrik   111
Stobbe, Tjebbe   137
Stork, Hijsch   128
Strong, John   22, 38
Sulzer Bros.   139
Suttorp, L. C.   55, 58, 62
Swaan, A.   77, 83, 84
Swart, P.   85, 87

Taurel, Louis J. M.   81, 82–91, 94, 96, 98, 101, 111, 147
Terrel   34, 35
Thompson, J. D. A.   17, 29, 30, 157, 158
Tjebbes, Klaas   155

Ueberfeld, Wed.   68

Veen, Johan van   51, 52
Visser, Jan Folkerts   40–42
Visser, J. T.   38
Visser, Pieter   69, 70
Volker, A.   137
Vreedenberg, A. C. J.   108
Vries, Jeronimo Catharinus de   107
Vries, Volkert de   51
Vrouwes, J.   98, 101

Wal, A. J. van der   105, 106
Wallen, P. van der   119–122
Weinholt, Arnold   39
Weinholt, Daniel   39
Wever, Cornelis   81
Whorlow, Wm   69
Wichers, P. J.   104
Wienen, W. van   137, 147, 148
Wijker, Aldert   82
Wijker, Gerrit   85, 87
Wijma, Sjoerd S.   65, 77
Wijsmüller   120, 137
Wilkens   121
William I   63, 66, 79, 153
William III   99
William V   26
Witsen, N.   50
Witsen, N. C.   102, 103
Witsen   98, 102, 103

167

Wood 34, 35
Wouter 106

Zaal, Van 128

Zeijlemaker, Jan 65, 68, 70
Zunderdorp, L. 42
Zurmühlen 116
Zuylen van Nyevelt, J. A. van 69

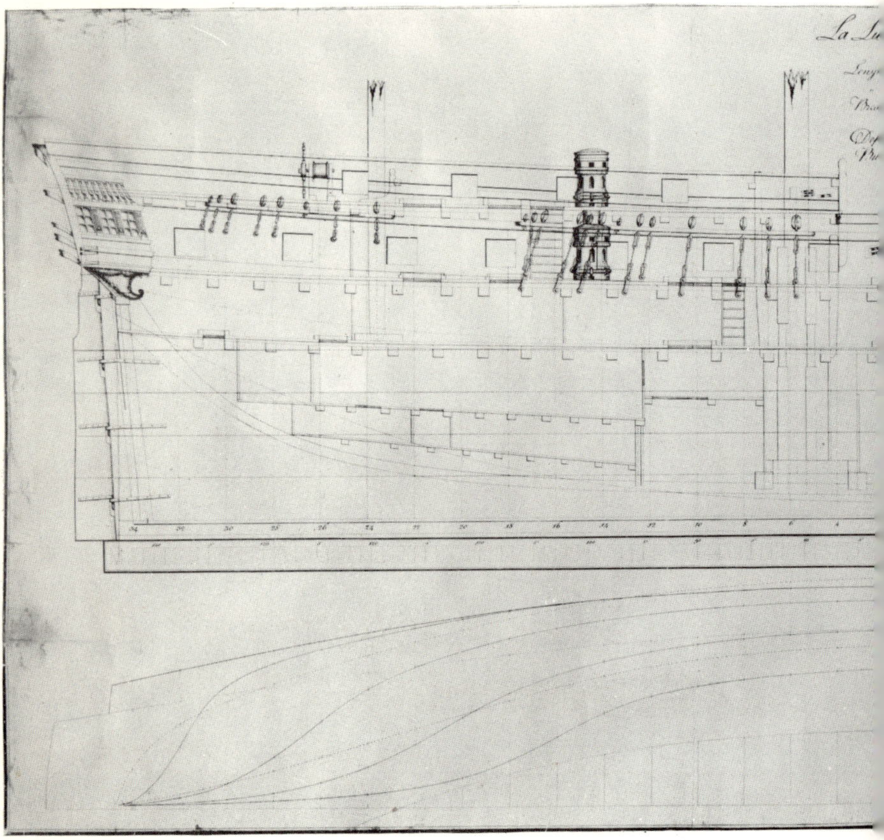